CONTRACTOR'S PORTABLE HANDBOOK

CONTRACTOR'S PORTABLE HANDBOOK

R. Dodge Woodson

McGraw-Hill

New York San Francisco Washington, D.C. Auckland Bogotá
Caracas Lisbon London Madrid Mexico City Milan
Montreal New Delhi San Juan Singapore
Sydney Tokyo Toronto

Library of Congress Cataloging-in-Publication Data
Woodson, R. Dodge (Roger Dodge), 1955–
   Contractor's portable handbook / R. Dodge Woodson.
      p.   cm.
   Includes index.
   ISBN 0-07-071836-9
   1. Building—Handbooks, manuals, etc. I. Title.
TH151.W57   1998
690'.837—dc21   97-40177
                                 CIP

McGraw-Hill

A Division of The McGraw·Hill Companies

1 2 3 4 5 6 7 8 9 0   DOC/DOC   9 0 3 2 1 0 9 8

ISBN 0-07-071836-9

The sponsoring editor for this book was Larry Hager, the editing supervisor was John C. Baker, and the production supervisor was Claire Stanley. It was set in Times Roman by Lisa M. Mellott through the services of Barry E. Brown (Broker—Editing, Design and Production).

Printed and bound by R. R. Donnelley & Sons Company.

McGraw-Hill books are available at special quantity discounts to use as premiums and sales promotions, or for use in corporate training programs. For more information, please write to the Director of Special Sales, McGraw-Hill, 11 West 19th Street, New York, NY 10011. Or contact your local bookstore.

 This book is printed on recycled, acid-free paper containing a minimum of 50% recycled, de-inked fiber.

This book is dedicated to Afton and Adam,
the two best kids a father could ever hope for.

CONTENTS

ACKNOWLEDGMENTS

I'd like to thank the U.S. Government for allowing some of their drawings to be used in this book as visual aids for readers. Additionally, I'd like to thank my parents for supporting my desire to make a career in the construction trades.

INTRODUCTION

Being a contractor in this fast-paced world can be quite a challenge. Times have changed, and contractors must keep up with the changes to remain competitive in their fields. More than ever, general contractors have to have knowledge in specialized fields, such as plumbing and electrical work. Customers ask a lot more questions than they did two decades ago. As the buying public becomes better informed, so must contractors.

How do you keep up with changes in the trades? Do you go to seminars? Many contractors do, but seminars are expensive and require a chunk of time out of at least one day. Because you are holding this book, you probably turn to the printed page to gain the knowledge you need. If this is the case, you are holding the right book.

My name is Roger (Dodge) Woodson. I've worked in the trades for more than 20 years and have taught classes at technical college. To this day, I'm active in remodeling and plumbing. At one time, I was building up to 60 houses a year. While I can't say I've seen and done it all, I've certainly witnessed more than most contractors. During my long career in construction, plumbing, and remodeling, I have learned a lot. I'd like to share some of my knowledge with you by having you read this book.

I'm licensed as a general contractor and a master plumber. When building homes for myself, I have installed the wiring and heating systems. After building dozens of houses year after year, remodeling countless homes, and running a full-service plumbing business, I've gathered some valuable information. I decided to compile a lot of this information into an easy-to-read handbook that you can use on a daily basis.

What do you want to know about? Are you interested in knowing more about roof framing and materials? Maybe your primary interest lies in HVAC systems. Whatever your interest, you will probably find answers to your questions in the following pages. The charts and tables that accompany the text are a big help in making this handbook both fast and easy to use.

As you read through the following chapters, you will discover a number of case histories from my past. Many of them deal with remodeling jobs. Experienced contractors know that much more is likely to go wrong on remodeling jobs than on new construction. Every chapter in this book gives you a key to new information on some aspect of construction, remodeling, and mechanical trades.

Take a few minutes to scan the table of contents. Turn to the back of this book, and look at all of the conversion tables, mathematical calculations, and other useful information that is available to you at a glance. Dig through the chapters, and read a little here and there. If I don't miss my guess, you will soon see this handbook as a must-have tool for your truck and office.

CHAPTER 1
SITE WORK

Site work is often part of a contractor's working life. Unfortunately, this is an area of construction that many contractors don't know enough about. Digging a trench in the wrong place can prove deadly. Failure to call utility companies prior to digging can be both dangerous and expensive. Fines for digging in an unapproved area can be steep. Are you aware of zoning ordinances? Do you know much about the safety requirements for digging around an existing home? Well, let's take a look at some of the key points pertaining to site work.

ZONING

Let's talk first about local zoning laws. These regulations can bring an abrupt halt to any job. Zoning regulations can govern everything from setbacks to the types of uses a building is approved for. For example, it might be in violation of zoning laws to convert a single-family home into a duplex. The same zoning laws can regulate how closely you can dig to a stream or property line.

If you are working in a progressive location, where code enforcement is done on a professional level, you will normally encounter any zoning problems prior to having a building permit issued. In some rural locations, however, things sometimes slip through the cracks until work has already started. This can be an expensive mess to straighten out. It should also be noted that your responsibility runs further than just finding out what zoning requirements entail. You must conform to the local laws, rules, and regulations as you proceed with your work.

Set-back Ordinances

Set-back ordinances are one of the most common zoning regulations with which contractors are faced. Whether you are building a garage, sun room, deck, or room addition, set-back requirements might alter your plans.

You might not feel that set-backs are your responsibility, but I wouldn't hang my hat on it in a court of law. As a professional contractor, you have certain obligations to your customers. It is not fair for me to say that you alone are to be held accountable for set-back violations, but if you don't want to spend the next few years in and out of court, you will do well to avoid conflicts of this type.

The issue of set-back requirements is not a complicated one. They are very simple, really. Most jurisdictions require a certain distance from every property line to any permanent structure, such as a house or garage. Portable structures, like sheds on skids, might or might not be affected by set-backs. Because every town, county, and state can create their own rules, you should check with your local authorities before commencing any work.

Set-backs vary in their rules. One town might say that front and back set-backs are 25 feet and that side set-backs are 15 feet. Another jurisdiction might set the front and back limits at 15 feet and the side set-backs at 10 feet. In the rural (or maybe you'd call it remote) setting where I live, there are no set-back requirements. The point is this: Set-back requirements do vary, and you should be informed of what each and every one of them dictate in areas where you will be working.

If you will be working in heavily populated areas, you are more likely to encounter trouble with set-backs than someone who is working out in the country. I have found this to be true of my career. Set-backs in Northern Virginia, where I worked for several years, were much more of an issue than they are here in Maine.

Anywhere where the price of land is at a premium, developers attempt to cram the land full. This means that homes and businesses are built on small lots and often are built right to the fringes of set-back requirements. Something as simple as installing a bay window could trigger a set-back problem. If the structure is in compliance with 1 foot of space to spare, and you add a bay window, you might throw the property into violation. This might seem unlikely to you, but I can tell you from experience that there are buildings that have been built to such strict guidelines. Don't take anything for granted.

Above-ground Structures

Above-ground structures, such as decks off of second floors, are not immune from set-back regulations. If you have a customer who wants a nice deck installed off the master bedroom on the second floor, you'd better check into zoning requirements. Even though no foundation work will be required, the deck will be an appurtenance to the building, and it might violate set-back laws. This type of work is almost always ignored by contractors in terms of zoning. Many contractor's assume that, because the foundation footprint of the building is remaining the same, there is nothing to worry about. Don't count on it!

Would you ever stop to think of set-back requirements when a customer asked you to install a large, garden window over a kitchen sink? I doubt it. If I hadn't been in the business for so long, I probably wouldn't think of it either. However, I have endured decades in the construction and remodeling business, and it has taught me a lot. Because a garden window will protrude further than a standard window, it might trigger a set-back violation. The odds of a house being this close to a property line is minimal, but it never hurts to check. It only hurts when you don't.

Free-standing Structures

Free-standing structures might offer some of the highest risks to a contractor. If you are called to build a free-standing garage, gazebo, or other outdoor structure, you must check the zoning laws carefully.

Whenever an outside structure is added to a property, it raises some risks for the contractor. Even if you are being hired only to build a barbecue pit, you should check the local zoning regulations. In short, if you are doing anything to alter the exterior of a property, zoning laws should be considered. However, it is not only outside work that could put you in a mess. Some inside work can cause just as much, if not more, trouble.

Inside Work

Inside work is not normally thought of as being a high risk in terms of zoning regulations, but it can be. Every now and then, a person buys a house or building with the intent to change its use. This is an extremely lucrative field of business when done properly, but a lot of rookie investors make mistakes. You might wind up right in the thick of things with them, if you don't know how to protect yourself.

Keep in mind that different locations, laws, regulations, and rules can affect the level of responsibility that a remodeling contractor has to any of the subjects we are discussing. Your part of the country might not require as much of you as some other parts. However, even if you are not held responsible for violations personally, the fact that violations are created can cut off your cash flow and force you out of business. This certainly brings the matter close to home and to the front of your attention.

LOCAL CODE REQUIREMENTS

Local code requirements are another issue that can put you at risk. How many contractors do you know who cheat the system? Certainly, you know plumbers who install water heaters without permits. I would guess you can name a few contractors who do interior work without permits. It is difficult for a contractor to embark on a major addition that requires site work without posting a permit, but it is sometimes done. This is a very bad practice, and it's one that can put you out of business quickly.

Most jurisdictions require permits for any substantial amount of work done on a property. If these permits are not obtained in the proper manner, the contractor is usually at risk. Anyone who has been in the business for very long has had customers who wanted to forgo the permit process. Homeowners know that taking out permits is paramount to be reassessed for higher property taxes. This is the homeowner's problem, not yours. However, if you go along with this approach and do the work without a permit, you become a party to the infraction.

It might be hard to believe that contractors would actually try to build a garage without a building permit, but some do. This is work that is out in the open and easily spotted by code officials. Yet, some contractors are so hungry for work that they will infringe on the laws and rules of the code enforcement office to get work. This is a shame, but it is also a big mistake. Contractors, in general, don't enjoy a great reputation. There are many stories about rip-offs and fraudulent activity centered around contractors, especially home-improvement and remodeling contractors. Don't get caught up in this situation. If a permit is required for the work you are bidding, don't do it without a permit.

DEED RESTRICTIONS AND COVENANTS

Deed restrictions and covenants are items that many contractors never think twice about. They aren't really a contractor's responsibility, but they can have a tremendous impact on your profits and stability. This is especially true in densely populated areas. Before you agree to do any exterior work on a building, you should seek some protection. Ask your customers to either provide you with a copy of their deeds or a release of liability for any violation of deed restrictions and covenants. A written release is your best course of action.

Land developers have the right to insert covenants and restrictions into the deeds of properties that they convey out to purchasers. There are no real limitations on what these restrictions may apply to. For example, a developer can prohibit the parking of commercial vehicles, such as your work truck, from being parked on the grounds of a property. Can you imagine living in a place where you couldn't park your truck in your own driveway? Well, my parents live in such a place. Their deed restricts not only the types of vehicles that can be kept on the property, but it goes on to dictate the type of mailbox they can have, the colors in which their house can be painted, and other types of situations that might arise. Being an old country boy, I can't accept this type of control, but I do understand it.

Developers put restrictions in their transfer deeds to protect their investments. A high-class subdivision could be demeaned if someone decided to paint their home pink, blue, and yellow. I suppose work trucks can also lend a less appealing atmosphere to a neighborhood than some fancy sports cars. From my personal perspective, this seems to be snobbish and wrong. However, from a developer's point of view, I understand the need for such regulations.

If people are going to pay top dollar for a prestigious address, they want to make sure their investment is protected. This, in many ways, is good for the developer and the property owners. While I understand it, and perhaps agree with some aspects of it, I feel it goes against the grain of what America stands for. In any event, the fact is that there are places where homeowners are limited in what they can do with the exterior appearance of their homes. This can affect you as a remodeling contractor. How does it bother you? Well, let's see.

Assume that you are awarded a job to install vinyl siding on a home. The customer has decided that routine painting is expensive and undesirable. For this reason, you have gotten the job to side the home with vinyl. It's a big house, and your profit margin is good. The only glitch in the deal is that the customer refuses to pay an up-front deposit. The reason you are given for this is the many stories that the homeowner has read about dishonest contractors.

The house is in a great neighborhood, and the homeowner seems fine. Because you have 30-day accounts for your materials and the job will take less than a week, you decide to accept the job offer. After all, the homeowner has agreed to pay you in full upon completion. The next week, you set up shop and begin the new siding installation. Your crew finishes the job on schedule, and you submit an invoice for payment. However, the payment doesn't come promptly. You call the customer and don't like what you hear.

When you contact the customer, you discover that the neighborhood homeowner's group has complained about having a house covered with vinyl siding. You never stopped to think that the house you were working on was the only house in the area with vinyl siding. The customer goes on to explain how the homeowner's board is bringing a legal suit to action that will force the removal of the new siding. In your position, you could care less about the lawsuit; you only want your money. However, the homeowner refuses to pay you until the dispute is settled. What are your options?

You can lien the house, but that won't automatically put money in your pocket. The supplier's bill for the materials will be coming soon, and you've already paid your crew. You are out money now, and you will have to pay your supplier by the tenth of the month. Still, you are getting nothing from the customer. Was it your responsibility to verify if your work was in compliance with subdivision restrictions? Probably not, but the result is the same: You're not getting paid. You could have avoided this situation if you had reviewed the customer's deed restrictions and covenants. Getting a release signed by the customer would give you more ammunition to go to court with, but you don't have any of this. All you've got are incoming bills and no cash. See how easy it is to wind up in a mess?

A New Roof

If you were called to install a new roof, would you consider the consequences for deed restrictions and covenants? You should. Roofing frequently falls into a category of control in subdivisions. There is often language in these restrictions that limits the types of roofing materials that can be used and the colors of roofing that are acceptable. Again, this is not an area where you should have to be on top of every deed restriction; however, if you're not, it can hurt you.

Landscaping

As a contractor, you might or might not take on landscaping jobs. (See Figures 1.1 through 1.3.) They can be a part of your overall project, or perhaps you do them for some extra income. In any case, landscaping can be affected by covenants and restrictions. If you are doing any work which affects the exterior of a property, you must be sure that the activity will not create problems that you cannot deal with effectively.

Name of tree	Recommended spacing (in feet)
Douglas Fir	12
White (paper) Birch	15
Quaking Aspen	7
White Spruce	10
Red Cedar	7
Eastern White Pine	12
Pin Oak	30
Sea Grape	4
American Holly	8
Lombardy Poplar	4
Oriental Arborvitae	3
Sugar Maple	40
Red Maple	25

FIGURE 1.1   Tree spacing.

Type of tree	Name of tree	Tree characteristics
Evergreen	Douglas Fir	Fast grower
Deciduous	White (paper) Birch	Very hardy
Deciduous	Quaking Aspen	Provides excellent visual screen
Evergreen	White Spruce	Structure breaks wind well
Evergreen	Red Cedar	Can grow well in dry soil
Evergreen	Eastern White Pine	Grows very fast
Deciduous	Pin Oak	Keeps leaves in winter
Evergreen	Sea Grape	Extremely decorative
Evergreen	American Holly	Beautiful leaves and berries
Deciduous	Lombardy Poplar	Grows fast and tall
Evergreen	Oriental Arborvitae	Grows fast
Deciduous	Sugar Maple	Beautiful tall foliage
Deciduous	Red Maple	Grows fast with great fall foliage

FIGURE 1.2 Tree characteristics.

Name of tree	Expected height (in feet)	Expected width (in feet)
Douglas Fir	60	25
White (paper) Birch	45	20
Quaking Aspen	35	5
White Spruce	45	20
Red Cedar	50	10
Eastern White Pine	70	40
Pin Oak	80	50
Sea Grape	20	8
American Holly	20	8
Lombardy Poplar	40	6
Oriental Arbovitae	16	6
Sugar Maple	80	50
Red Maple	40	30

FIGURE 1.3 Tree specifications.

GRADING

Grading work can get expensive very quickly. If your work will alter existing conditions or create new circumstances where existing grading will not be sufficient, you have to be aware of this. People can get pretty nasty at times, and the only way to avoid confrontations and lawsuits is to address the issues and objections before they become volatile. Let me give you an example of such a situation.

I worked for a contractor once who was hired to build a sun room on the back of a house. The house was on a sloping lot, and the addition's foundation was going to extend further above the finished grade than the rest of the home's foundation. This issue was not discussed prior to the construction process. My supervisor was aware of the increased foundation exposure, but he failed to discuss it with the homeowners. I don't know all of the particulars, but I can fill you in on the key points of the problem.

A new sun room was built on an existing home. When the foundation was installed, which was made of cinder block, it extended well above grade. The customers were under the impression, for whatever reasons, that the foundation would be backfilled to match up with their existing foundation. The contractor didn't share their opinion. One thing led to another, and a major battle ensued. The property owners wanted the contractor to haul in dirt to conceal the ugly block foundation. My boss had no intentions of doing this, and they all wound up in court. I don't know what the outcome of the dispute was, because I changed jobs shortly after the trouble roared into full force. This was a problem in communications, but it was a serious problem, nonetheless.

ENVIRONMENTAL CONCERNS

Environmental concerns have become an issue for remodelers to be mindful of. Excavating for a new foundation or filling in some low spot on a lot can get you in trouble from an environmental angle. There are state and federal laws pertaining to what can and cannot be done in certain areas. For example, here in Maine, wetlands is a big issue.

Maine has a lot of land that is considered to be wetland. This wetland is heavily protected from development. Building too close to a designated wetland or filling in such a piece of property is a major offense. This type of situation can affect a contractor.

Let's say that you are called out to build a terraced deck for a customer. The deck will start at the rear of the home and extend downward, towards some lake frontage. Your customer wants this deck to enjoy sunsets over the water. After looking over the proposed location for the deck, you agree to build it. The furthest end of the deck stops well short of the lake, but it terminates in some reeds where the land begins to get soft. If you build this deck, you are most likely violating environmental laws.

In the scenario we've just examined, you might be saved from big trouble by having your building permit request denied. However, maybe the permit would slip through. This gives you your right to build the deck; however, if you construct it in a protected area, you might still be held accountable. At the least, you will have to defend your actions, and this will eat up whatever profits you might have made. There are all sorts of tricky situations like this one that can complicate your life.

If you will be doing site work in an area where erosion or runoff might occur, you might very well be required to install a barrier to prevent such occurrences. The barrier might be made of bales of straw or plastic. If you are unaware of this requirement when you bid a job, the cost of this work will come out of what you had hoped to be profit. You're in business to make money, not lose it, so do your homework before you commit to a quoted price.

Even if you believe responsibility for site considerations does not rest on your shoulders, you should protect yourself at all times. (See Figures 1.4 through 1.8.) Being issued a building permit should take you off the hook, but fighting in court to prove that you are innocent gets very expensive. Believing that your customers are responsible for what they hire you to do is not good enough. Ignorance of the law is not a suitable defense. As a professional contractor, it is up to you to know all the laws pertaining to your actions and to avoid infractions of them.

Soil type	Drainage rating	Frost heave potential	Expansion potential
Bedrock	Poor	Low	Low
Well-graded gravels	Good	Low	Low
Poorly graded gravels	Good	Low	Low
Well-graded sand	Good	Low	Low
Poorly graded sand	Good	Low	Low
Silty gravel	Good	Moderate	Low
Silty sand	Good	Moderate	Low
Clayey gravels	Moderate	Moderate	Low
Clayey sands	Moderate	Moderate	Low

FIGURE 1.4   Soil properties.

Type of soil	Percent of swell	Percent of shrinkage
Sand	14–16	12–14
Gravel	14-16	12–14
Loam	20	17
Common earth	25	20
Dense clay	33	25
Solid rock	50–75	0

FIGURE 1.5 Swelling and shrinkage of soils.

Material	Support capability
Hard rock	80 tons
Loose rock	20 tons
Hardpan	10 tons
Gravel	6 tons
Coarse dry sand	3 tons
Hard clay	4 tons
Fine dry sand	3 tons
Mixed sand and clay	2 tons
Wet sand	2 tons
Firm clay	2 tons
Soft clay	1 ton

FIGURE 1.6 General bearing capacities of foundation soils.

UNDERGROUND UTILITIES

Underground utilities are common in many locations. If you start digging footings with a backhoe, without proper planning, you might wind up either in big trouble or dead. Digging up a buried gas pipe or electrical cable can bring your career to a quick end.

There are usually agencies that you can call to help avoid underground mishaps. If there is no single agency in your area that handles all underground utilities, you should call each utility company that might have cables, pipes, or equipment in the area where you will be working. The agencies and companies will come to the site and mark the locations of their underground materials. There is still a chance you might find something buried where it is not supposed to be, so caution should still be observed. However, if you have had

Tons/sq. ft. of footing	Type of soil
1	Soft Clay Sandy loam Firm clay/sand Loose fine sand
2	Hard clay Compact fine sand
3	Sand/gravel Loose coarse sand
4	Compact coarse sand Loose gravel
6	Gravel Compact sand/gravel
8	Soft rock
10	Very compact Gravel and sand
15	Hard pan Hard shale Sandstone
25	Medium hard rock
40	Sound hard rock
100	Bedrock Granite Gneiss

FIGURE 1.7   Safe loads by soil type.

Trench depth	Trench width
1 foot	16 inches
2 feet	17 inches
3 feet	18 inches
4 feet	20 inches
5 feet	22 inches
6 feet*	24 inches

*__Note:__ Trenches with depths of more than five feet should be shored up or fitted with a sheath for personal safety.

FIGURE 1.8   Suggested trench depths and widths.

all of the utility locations marked by proper authorities, you have at least removed yourself from any blame for negligence.

I've talked to a lot of general contractors in the past who don't worry about underground utilities. Their position is that they hire subcontractors to do their digging, so the problem rests with the subs, not with the generals. In theory, this might be true. However, it doesn't take a lot of time or effort to make a few phone calls, and I always prefer to know that the calls have been made. For this reason, I insist on calling the utility agencies myself. If the diggers call too, that's fine, but I want to know that my office went on record for requesting underground utility markings.

TREES

It is not uncommon for trees to be in the way of large remodeling projects. This could be the case if you are raising a roof or adding an addition. Who is going to be responsible for removing the trees? If you're bidding the job with an open, vague proposal, the customer will probably assume that you are taking care of the tree removal. If you aren't, you had better spell it out in your proposal.

Cutting trees can be dangerous under any conditions, and the risk goes up when the trees are near homes and buildings. Professional tree cutters don't work cheap, but they are a good investment under most remodeling circumstances. You could send a laborer to the job with a chainsaw to hack down the tree, but what will you do if the tree falls on a house or car? Worse yet, what happens if the tree falls on a person? In my opinion, the risk of cutting trees in populated areas is high enough to warrant the expense of a professional tree cutter who is properly insured.

DAMAGED LAWNS

I have heard various contractors complain about how their jobs turned into nightmares because of damaged lawns. In general, it seems that their customers were not prepared for the type of destruction that some jobs can cause to a lawn. When a backhoe rolls across a lush lawn to dig a footing, leaving deep tread marks in its path, unprepared customers have a right to get upset. The customers are not experienced contractors. They don't think about the fact that equipment has to get to the job site in some way, which often means right across their lawn. Neither do they stop to think about how loose nails will wind up in the lawn when a new roof is installed.

As a business owner, you owe it to your customers to advise them of situations that might arise around your work. If you're going to have to dig up their yard to make a new sewer connection for an addition, tell them what to expect. If you will be setting up pump jacks or staging that might kill some of their grass, discuss it with the customer. Communication goes a long way in avoiding confrontations.

BE THOROUGH

Be thorough in your site inspections. Look for circumstances that might lead to unexpected expenses. Check to see if there is likely to be anything happening that the customer hasn't thought of. If you don't do your own site work, take your site contractor with you on the inspection.

After you have looked the job over carefully, check with local authorities to determine if there are any hidden problems with the work that the customer is requesting. Talk to the utility companies. If there is a gas main running underground where your customer wants footings dug for a new garage, something in the plan is going to have to be changed.

Cover all of your bases before you commit to a firm price. This is the only way to be fair to yourself and your customers.

CHAPTER 2
PRIVATE WATER SUPPLIES

Site evaluations for water wells are often taken for granted by builders. I have known many builders who submitted bids for work without ever seeing the building lot. There is a lot of risk in doing this. A builder cannot bid a job competitively without knowing what type of well will be used. If a bid goes in for a bored well and it turns out that a drilled well is needed, the bidder will lose money. When compensation for a site inspection is made by bidding a job with the most expensive type of well, the job can be lost by being bid too high. Failure to do a site inspection can be a very big mistake.

You can't always see what's likely to be under the ground by looking at the surface. Knowledgeable builders want to know what they will get into when drilling wells and digging footings. Many experienced builders require customers or land owners to provide them with soil studies before giving a firm bid. When such studies have not been done, some builders do their own. I'm one of these builders.

It is not uncommon to see me out digging holes on potential building sites. A post-hole digger can reveal a lot about what conditions exist below the top soil. Augers and probe rods can also provide some insight into what is likely to be encountered. A probe rod will tell you if a lot of rock is present. However, to see the soils, you need a hole. An auger or post-hole digger is the best way to get these samples. Augers are often easier to use, and a power auger is ideal.

When you create test holes, you have a lot more information to base your bid decisions on. There are only a few ways to bid a job where a well will be installed. You can guess what will be needed, but this is very risky. Digging test holes will give you a very good idea of what types of wells might be suitable. Interviewing well owners on surrounding property can provide a lot of data that can help you with your decision. Hiring soil-testing companies is a great, but expensive, way to find out what you are getting into. Having a few well installers walk the land with you so that they can provide you with solid bids is another good way to protect yourself.

LOCATION

The location of a well is important. Choosing a location is not always easy. Many factors can influence the location of a well. The most obvious might be the location of a house. It is not common to place a well beneath a home, so most people will choose a location outside of the foundation area of a home. Septic fields are another prime concern. Wells are required to be kept a certain distance from septic fields. The distance can vary from jurisdiction to jurisdiction and because of topography. Access to a location with a well-drilling rig is also a big factor. These large trucks aren't as maneuverable as a pick-up truck. Picking a place for a well must be done with access in mind.

Finding suitable sites for a well can be made easier with some background information and site inspections. Maps can give you a lot of guidance on where water might be found. Some regional authorities maintain records on wells already in existence. Reviewing this historical data can definitely help you pinpoint your well type and location.

Unfortunately, there is never any guarantee that water will be where you think it is. A neighboring landowner might have a well that is 75 feet deep, while your well turns out to be 150 deep. It is, however, likely that wells drilled in close proximity will average about the same depth. I've seen houses in subdivisions where one house has great well water and the next-door neighbors water suffers from unpleasant sulfur content. Fifty feet can make quite a difference in the depth, quantity, and quality of a well.

Plants

Plants can be very good indicators of what water is available beneath the ground's surface. Trees and plants require water. The fact that plants and trees need water might not seem to provide much insight into underground water. Take cattails as an example. If you see cattails growing nearby, you can count on water being close by. It's suggested that the depth of water in the earth can be predicted, to some extent, by the types of trees and plants in the area.

Cane and reeds are believed to indicate that water is within 10 feet of the ground's surface. Arrow weed means that water is within 20 feet of the surface. There are many other references to various types of plants and trees in regard to finding water. From a well-drilling point of view, I'm not sure how accurate these predictions are. I know that cattails and ferns indicate water is close by, but I can't say that it will be potable water or how deep it will be found in the ground. I suspect that there are some very good ways to predict water with plants and trees.

Because I am not an expert in plants, trees, or finding water, I won't attempt to pump you up with ideas of how to find water at a certain depth just because some particular plant grows in the area. It is my belief that, with enough knowledge and research, a person can probably predict water depths with good accuracy, in many cases.

DRILLED WELLS

Drilled wells are the most dependable individual water source I know of. These wells extend deep into the earth. They reach water sources that other types of wells can't come close to tapping. Because drilled wells take advantage of water that is found deep in the ground, it is very unusual for drilled wells to run dry. This accounts for their dependability. However, dependability can be expensive.

Of all the common well types, drilled wells are the most expensive to install. The difference in price between a dug well and a drilled well can be thousands of dollars. However, the money is usually well spent. Dug wells can dry up during hot, summer months. Contamination of well water is also more likely in a dug well. When all the factors are weighed, drilled wells are worth their price.

Depth

The depth of a drilled well can vary a great deal. In my experience, drilled wells are usually at least 100 feet deep. Some drilled wells extend 500 feet or more into the earth. My personal well is a little over 400 feet deep. Based on my experience as both a builder and plumber, I've found most drilled wells to range between 125 to 250 feet deep. When you think about it, that's pretty deep.

It's hard for some people to envision drilling several hundred feet into the earth. I've had many home buyers ask me how I was going to give them such a deep well. Many people have asked me what will happen if the well driller hits rock. Getting through bedrock is not a problem for the right well-drilling rig. While we are on the subject of drilling rigs, let's talk about the different ways to drill a well.

Well-drilling Equipment

Well-drilling equipment is available in various forms. While one type of rig might be the most common, all types of well rigs have advantages and disadvantages. As a builder, it can be helpful to know what your options are for drilling wells on different types of sites. There are two basic types of drilling equipment.

Rotary drilling equipment is very common in my area. This type of rig uses a bit to auger its way into the earth. The bit is attached to a drill pipe. Extra lengths of pipe can be added as the bit cuts deeper into the ground. The well hole is constantly cleaned out using air, water, or mud under pressure.

Percussion cable tool rigs make up the second type of drilling rig. These drilling machines use a bit that is attached to a wire cable. The cable is raised and dropped repeatedly to create a hole. A bailer is used to remove debris from the hole. Between rotary equipment and percussion cable equipment, there are a number of variations in the specific types of drilling rigs.

We could go into a lengthy discussion of all the various types of well rigs available. However, because you are a builder and probably have no desire to become a well driller, there seems to be little point in delving into all the details of drilling a well. What you should know, however, is that there are several types of drilling rigs in existence. Your regional location might affect the types of rigs being used. A few phone calls to professional drillers will make you aware of what your well options are. For my money, I've also contracted the services of rotary drillers. There are times when other types of rigs could be a better choice. My best advice to you is to check with a number of drilling companies in your area and see what they recommend.

THE BASICS

Let's go over the basics of what is involved, from a builder's point of view, when it comes to drilling a well. Your first step must be deciding on a well location. The site of a proposed house will, of course, have some bearing on where you want the well to be drilled. Local code requirements will address issues pertaining to water wells. For example, the well will have to be kept at some minimum distance from a septic field, assuming that one is to be used.

Who should decide on where a well will go? Once local code requirements are observed, the decision for a well location can be made by a builder, a home buyer, a well driller, or just about anyone else. If you're building spec houses, the decision will be up to you and your driller. Buyers of custom homes might want to take an active role in choosing a suitable well site. As an experienced builder, I recommend that you consult with your customers on where they would like their well. Some people are adamant about where they do and do not want a well placed.

Access

Access is one of the biggest concerns a builder has in the well-drilling process. It is a builder's responsibility to make access available to a well driller. Drilling rigs require a lot of room to maneuver. A narrow, private drive with overhanging trees might not be suitable for a well rig.

Having enough width and height to get a well truck into a location is not the only consideration. Well rigs are heavy—very heavy, in fact. The ground that these rigs drive over must be solid. New construction usually requires building roads or driveways. If you can arrange to have a well installed along the roadway, your problems with access are reduced.

It is not always desirable to install a well alongside a driveway. This may not cause any additional concern, but it could. If the ground where you are working is dry and solid, a well rig can drive right over it. However, if the ground is wet and muddy, or too sandy, a big truck won't be able to cross it. You must consider this possibility when planning a well installation.

Don't build obstacles for yourself. Just as the old joke goes about painting yourself into a corner, you can build yourself right out of room. Some builders put wells in before they build houses. Others wait until the last minute to install a well. Why do they wait? They do it to avoid spending the money for a well before it's needed. This reduces the interest they pay on construction loans and keeps their operating capital as high as possible. When the well is installed last, it can often be paid for out of the closing proceeds from the sale of a house.

I have frequently waited until the end of a job to install wells. There is some risk to this method. It's possible, I suppose, that a house could be built on land where no water could be found with a well. This would truly be a mess. A more likely risk is that the house construction will block the path of a well rig.

When you are confirming the location for a well, make sure that you will be able to get drilling equipment to it when you need to. Going to the buyers of a custom home and informing them that their beautiful shade trees will have to be cut down to get a well rig into the site is not a job I would enjoy having. It is safer to install wells before foundations are put in.

Working on Your Site

Once you have a well driller working on your site, there's not much for you to do but wait. The drilling process can sometimes be completed in a single day. There are times, however, when the rig will be working for longer periods of time. Depending on the type of drilling rig being used, a pile of debris will be left behind. This doesn't usually amount to much, and your site contractor can take care of the pile when preparing for finish grading.

The well driller will usually drill a hole that is suitable for a 6-inch steel casing. The casing will be installed to whatever depth is necessary. Once bedrock is penetrated, the rock becomes the well casing. How much casing is needed will affect the price of your well. Obviously, the less casing that is needed, the lower the price should be.

Your well driller will grout the well casing as needed to prevent ground water from running through the casing and dropping down the well. This is normally a code requirement. It limits the risk of contamination in the well from surface water. A metal cap is then installed on the top edge of the casing, and then the well driller's work is done.

Pump Installers

Many well drillers offer their services as pump installers. You might prefer to have the driller install the pump, or maybe you would rather have your plumber do the job. A license might be needed in your area for pump installations. Any master plumber will be allowed to make a pump installation, but you should check to see if drillers are required to have special installation licenses. If they are, make sure that any driller you allow to install a pump is properly licensed.

If you want your driller to install a pump setup as soon as the drilling is done, you will need to make arrangements for a trench to run the water service pipe in. The cost of this trench is usually not included in prices quoted by pump installers. Watch out for this one, because it is an easy way to lose a few hundred dollars if you don't plan on the cost of the trench.

After your well is installed, you should take some steps to protect it during construction. Heavy equipment could run into the well casing and damage it. I suggest that you surround the well area with some highly visible barrier. Colored warning tape works well, and it can be supported with nothing more than some tree branches stuck in the ground. A lot of builders don't take this safety precaution, but they should. A bulldozer can really do a number on a well, and a well casing can be difficult for an operator to see at times.

How far will the pump installer take the job? Will the water service pipe be run just inside the home's foundation and left for future hook up? Who will run the electrical wires and make the electrical connections? Does the installation price include a pressure tank and all the accessories needed to trim it out? Who will install the pressure tank? You need answers to these questions before you award a job to a subcontractor. Pump systems involve a lot of steps and materials. It's easy for a contractor to come in with a low bid by shaving off some of the work responsibilities in fine print. Be careful, or you might wind up paying a lot more than you planned to for a well system.

QUANTITY AND QUALITY

When it comes to the quantity and quality of water produced by a well, few (if any) well drillers will make commitments. Every well driller I've ever talked to has refused to guarantee the quantity or quality of water. The only guarantee that I've been able to solicit has been one of hitting water. This is an issue that builders have to be aware of.

Customers might ask you to specify what the flow rate of their new well will be. It is beyond the ability of a builder to do this. An average, acceptable well could have a 3-gallon-per-minute recovery rate. Another well might replenish itself at a rate of 5 gallons per minute. There are wells with much faster recovery rates, and there are those with slower rates. Anything less than 3 gallons per minute is less than desirable, but it can be made to suffice.

How can you deal with the recovery rate? You really can't. Sometimes by going deeper, a well driller can hit a better aquifer that will produce a higher rate of recovery, but there is no guarantee. So, don't make a guarantee to your customers. Show your customers the disclaimers on the quotes from well drillers, and use that evidence to back up your point that there is no guarantee of recovery rate in the well business. Now, I could be wrong. You might find some driller who will guarantee a rate, but I never have.

Even though you can't know what a recovery rate will be when you start to drill a well, you can determine what it is after water has been hit. Your well driller should be willing to test for and establish the recovery rate for you. Every driller I've ever used as performed this service. You need to know what the recovery rate is in order to size a pump properly. There is an old-fashioned way of determining a recovery rate, but I'd ask my driller to do it for me.

It is also important that you know the depth of the well, and this is something that your driller can certainly tell you. The depth of the well will also be a factor when selecting and installing a pump. Don't let your driller leave the job until you know what the depth and the recovery rate is.

Quality

The quality of water is difficult to determine when a new well is first drilled. It can take days, or even weeks, for the water in a new well to assume its posture. In other words, the water you test today might offer very different results when tested two weeks from now. Before a true test of water quality can be conducted, it is often necessary to disinfect a new well. Many local codes require disinfection before testing for quality.

Wells are usually treated with chlorine bleach to disinfect them. Local requirements on disinfection vary, so I won't attempt to tell you exactly what to do. In general, a prescribed amount of bleach is poured into a well. It is allowed to sit for some specific amount of time, as regulated by local authorities. Then the well pump is run to deplete the water supply in the well. As a rule of thumb, the pump is run until there is no trace of chlorine odor in the tap water. When the well replenishes itself, the new water in the well should be ready for testing. Again, check with your local authorities for the correct procedure to use in your areas. (By the way, builders usually are the people responsible for conducting the disinfection process.)

Once the well is ready to test, a water sample is taken from some faucet in the house. Test bottles and collection instructions are available from independent laboratories. Follow the instructions provided by the testing facility. As a rule of thumb, you should remove any aerator that might be installed on the spout of a faucet before taking your water collection. It is often recommended that a flame be held to the spout to kill bacteria that may be clinging to the faucet and washed into the collection bottle. (Don't attempt this step when testing from a plastic faucet!) Many plumbers take water tests from outside hose bibbs, and this is fine. A torch can be used to sterilize the end of the hose bibb, and a hose bibb provides almost direct access to water.

When collecting water for a test, water should be run through the faucet for several minutes before catching any for a test. It is best to drain the contents of a pressure tank and have it refill with fresh well water for the test. You can run reserve water out quickly by opening the cold water faucet for a bathtub.

Once a water sample is collected, it is taken or mailed to a lab. Time is of the essence when testing for bacteria, so don't let a water bottle ride around in your truck for a few days before you get around to mailing it. Again, follow the instructions provided by the lab for delivering the water.

There are different types of tests that you can request the lab to perform. A mandatory test will reveal if the water is safe to drink. If you want to know more, you have to ask for additional testing. For example, you might want to have a test done to see if radon is present. Many wells have mineral contents in sufficient quantity to affect the water quality. Acid levels in the well might be too high. The water could be considered hard, which can make washing with soaps difficult and can leave staining in plumbing fixtures. There are a number of potential tests to run.

Drinking water, or *potable water* as it can be called, is your primary goal when drilling a well. It is rare that a drilled well will not test well enough to meet minimum requirements for safe drinking. In fact, I've never known of a drilled well that wasn't suitable for drinking from. Nevertheless, an official statement of acceptability is generally required by code officials and lenders who loan money on houses.

If you get into a discussion on water quality, you might have to pinpoint exactly what it is that you are talking about. Is the subject only related to the water being safe to drink? Or does it extend into mineral contents and such? Most professionals look at water quality on an overall basis, which includes mineral contents.

TRENCHING

Trenching will be needed for the installation of a water service and for the electrical wires running out to a submersible pump. It is possible to pump water from a deep well with a two-pipe jet pump, but submersible pumps are, in my opinion, far superior. When a submersible pump is used, it hangs in the well water. Electrical wires must be run to the well casing and down into the well. The wires and the water service pipe can share the same trench.

As with any digging, you must make sure that there are no underground utilities in the path of your excavation. Most communities offer some type of underground utility identification service. In many places, one phone call will be all that is needed to get all underground utility locations on your work site marked. It might, however, be necessary in some parts of the country to call individual utility companies. I expect that you are familiar with this process; most builders are.

Once you have a clear path to dig, a trench must be dug to a depth that is below the local frost line. The water service pipe will have water in it at all times, so it must be buried deep enough to avoid freezing. How deep is deep enough? It varies from place to place. In Maine, I have to get down to a depth of 4 feet to be safe from freezing. In Virginia, the frost line was set at 18 inches. Your local code office can tell you what the prescribed depth is in your area.

After a trench is opened up, you can arrange for your pump installer. Most installers will want to do all of their work in one trip. This means that you must have enough of the house built to allow an installer to bring the water pipe through the foundation and to set the pressure tank. As a helpful hint, you should install a sleeve in your foundation as it is being poured so that the pump installer will not have to cut a hole through the foundation.

Check with your local plumbing inspector to determine what size sleeve will be needed. Most plumbing codes require a sleeve in a foundation wall to be at least two pipe sizes larger than the pipe being installed. For a typical 1-inch well pipe, this would mean that the sleeve would have to be at least 2 inches in diameter. Again, check with your local code office, because plumbing codes do vary from place to place.

CHAPTER 3
INSTALLING A
PUMP SYSTEM

Installing a pump system for a drilled well is not a difficult procedure. It is, however, a job that is usually required to be done by a licensed professional. Even if you know how to do all of the work, you probably can't do it legally without a license.

Because you are a builder, rather than a pump installer, I will not go into every little detail of a pump installation. I will cover all the key points, but without the step-by-step instructions that I would give someone wanting to learn the trade. In other words, I'm not going to waste your time with instructions to apply pipe dope, how tight to turn a fitting, or how to carry out every other little plumbing process. What I will tell you is how to make sure that the installer you choose is doing a good job.

AT THE WELL CASING

Let's start our work at the well casing. You should be looking at an empty trench and the side of a steel well casing when a pump installation begins. A hole has to be cut through the side of the well casing to allow what is called a *pitless adapter* to be installed. A cutting torch can be used to make this hole, but most installers use a metal-cutting hole saw and a drill. The size of the hole needed is determined by the pitless adapter. Keeping the hole at proper tolerances is important. If the pitless adapter doesn't fit well, ground water might leak into the well by getting past the gasket provided with the pitless adapter.

The hole in the side of the well casing should be positioned so that a water pipe laying in the trench will line up with the pitless adapter once it is installed. Once the hole has been cut, the pitless adapter needs to be installed. This is really a two-person job. One person will work down in the trench, while another person will work from above the main opening in the well casing.

A long, threaded pipe is needed to position the piece of the pitless adapter that is installed inside the well. Most plumbers make up a T-shaped pipe tool to use for this part of the job. The pitless adapter is screwed onto the threads of the T-shaped tool. With this done, one person lowers the pitless adapter down the well and into position. A threaded protrusion on the adapter is intended to poke through the hole in the side of the well casing. When it does, the person in the trench installs a gasket and the other part of the pitless adapter, which is basically a retaining ring. The ring is tighten to create a watertight seal and to hold the pitless adapter in place. At this point, the T-shaped tool is unscrewed from the pitless adapter and laid aside. As a builder, you should check to see that the pitless adapter is tight and in position to prevent contaminated ground water from entering around the hole.

Most installers put their pump rigs together next. This involves one of two things: either a truck that is equipped with a reel system for well pipe or enough room to lay all of the well pipe out in a fairly straight line. With deep wells, this can be a problem. It is important, however, that the well pipe be laid out and not allowed to kink, unless it is being fed into the well from a reel system.

The pipe used for most wells comes in long coils. It is best to avoid joints in well pipe whenever possible. With the long lengths of pipe available, there should be no reason why a joint would be needed to splice two pieces of pipe together. When joints and connections are made with standard well pipe, it is best to use metallic fittings. Nylon fittings are available, but they might break under stress more quickly than metal fittings.

Polyethylene (PE) pipe is the type used most often for well installations. This black plastic pipe is sold in coils and can be used for both the vertical drop pipe in the well and the horizontal water service in the trench. In houses where hot water will be available, PE pipe must not extend more than 5 feet inside the foundation wall. The plumbing code requires the same pipe type to be used for water distribution of hot and cold water. PE pipe is not rated to handle hot water, so it cannot be used as an interior water distribution pipe.

Other types of pipe can be used for well systems, but PE pipe remains the most popular. This pipe has some drawbacks for installers. The material will kink easily. If the pipe kinks, it should not be used. The bend in the pipe weakens it. If you see an installer kink a pipe, make sure that the section of pipe that was kinked is not used. Because couplings are undesirable both in the well and in the trench, a kinked pipe might mean getting a whole new roll of pipe, depending upon where the kink occurs.

Another fault of PE pipe is its tendency to become very hard to work with in cold weather. Fittings, which are an insert type, are difficult to push into the pipe when it's cold. Warming the ends of the pipe with a torch or heat gun will make the material pliable and easy to work with. However, care must be used to avoid melting the pipe. All connections made with PE pipe should be made with two stainless-steel hose clamps. One clamp is all that is required by code, but a second clamp provides cheap insurance against leaks.

When a coil of PE pipe is unrolled, it will take some work to straighten it out. This is basically a two-person job, although I have done it alone. The pipe should be stretched out as straight as possible, without kinking it, and then the pipe must be manipulated in a looping motion to make it lay flat. This step of the installation process should not be ignored.

Once the well pipe is laying flat, an installer can proceed with putting the pump and accessories together with the pipe. A torque arrestor should be installed on the pipe. This will limit vibration in the well as the pump runs. A male, insert adapter (brass, please) should be used to connect the pipe to the pump. Another insert adapter will connect the pipe to the pitless adapter fitting that has yet to be installed.

Electrical wires must run from the pump to the top of the well casing. Some plumbers use electrical tape to secure electrical wire to the well pipe. This is a common procedure. Other installers use plastic guides that slide over the well pipe to secure the wires. Either way, someone has to make sure that the wires are secure. It is also very important to make sure that the wires are not damaged as the pump assembly is lowered into the well. Waterproof splice kits can be used to joint electrical wires that are extending into a well, and of course, waterproof wire is required.

All of the piping and wiring is connected to a submersible pump before the pump is installed. This work is normally done near the well, on the ground. Once the entire assembly is put together, it is lowered into the well. Some companies have special trucks for this part of the job, but a lot of installers do it the old-fashioned way: by hand.

When the pump assembly is put down into the well, the work goes much better if two people are involved. Lowering a pump assembly by hand is not difficult. The T-bar tool is once again connected to the part of a pitless adapter that is wedge-shaped and attached to the upper end of the drop pipe. Smart installers attach nylon rope to the submersible pump as a safety rope. I've seen a number of installers skip this step, but I won't allow it on my jobs. I insist on a safety rope being installed.

Submersible pumps are held in the well by only the well pipe, unless a safety rope is used. If a fitting pulls lose, an expensive pump can be lost in the well forever. The rope, which is ultimately tied to the top of the well casing gives you something to retrieve the pump with if anything goes wrong with the piping arrangement. For the low cost of nylon rope, it's senseless not to use it.

The pump is lowered into the well. During this process, the person working the assembly down into the casing must take care not to scrape the electrical wires along the edge or sides of the casing. If the insulation on the wires is cut by the casing, the pump might fail to operate.

Once the pump is in position, the pitless adapter piece is put into place with the part of the pitless adapter that is secured to the well casing. This is done simply by lining the wedge up with the groove in the stationary piece and tapping it into place. Then the safety rope is adjusted and tied off to the casing. The T-bar is unscrewed and removed. The only step remaining at the well head is the connection of electrical wires and the replacement of the well cap. An experienced crew can put together and install a pump assembly in about two hours, or so.

THE WATER SERVICE

The next step in the installation of a pump system is the water service. This pipe might be the same type as is used in the well, or it can be some other type. Copper tubing was used as a water service material for years. It is still used at times. Most installers use PE pipe, and there is no reason not to. It's less expensive than copper, it's not affected by acidic water in the same way that copper is, and PE pipe has a long life span.

PE pipe for a water service is prepared in the same way that it would be for use as a drop pipe. It is laid out and straighten to get the loops out of it. When this is complete, it is placed in the trench. One end of the pipe connects to the protrusion from the pitless adapter. This connection is made with two hose clamps. The other end of the pipe is placed through a foundation sleeve and extended into a home. The entire length of the pipe should be lying flat on the bottom of the trench. There should not be any rocks or other sharp objects under or around the pipe. This is important, and it is something that not all installers are really careful about, so you might want to check it yourself. Any sharp objects might puncture the pipe during the backfilling process. It is also possible for rough or sharp objects to the cut the pipe months after installation as the ground settles and the pipe moves. Make sure the trench and the backfill material are free of objects that might harm the pipe.

Water service pipes often have to be inspected before they are covered up. This is not directly your responsibility, but you should make sure that an inspection has been approved if one is required. Backfilling the trench should be done gradually. If someone takes a backhoe and pushes large piles of dirt into the ditch, the pipe might kink or collapse. Require the backfilling to be done in layers, so that there will not be excessive weight dumped on the pipe all at once.

INSIDE THE FOUNDATION

The rest of the work will take place inside the foundation. A pressure tank should be used on all jobs. This tank gives a reserve supply of water that allows the house to have good water pressure. The tank also preserves the life of the pump. Without a pressure tank, a water pump would have to cut on every time a faucet was turned on. This short cycling of the pump would wear it out quickly. A pressure tank removes the need for a pump to cut on every time water is needed. Until the tank drops to a certain pressure, the pump is not required to run. When the pump refills the tank, it runs long enough to avoid short cycling.

The size of a pressure tank is normally determined by the number of people and plumbing fixtures expected in a house. Larger tanks require the pump to run less often. In addition to the pressure tank, there are many accessories that must be installed. For example, there will be a pressure gauge, a relief valve, a drain valve, a pressure switch, and a tank tee. Electrical regulations might call for a disconnect box at the tank location. Even if your local code doesn't require a disconnect box, it's a good idea to install one.

Under normal circumstances, an experienced installation crew can complete an entire pump installation, including inside work, in less than a day. As a builder, you will have to hang around the job if you want to see that all of the work is being done correctly, because it will happen quickly. However, you can tell a lot about the workmanship by making periodic inspections.

PUMP SIZING

Pumps are rated on their output in gallons per minute (GPM). Let's say that you have a well with a recovery rate of 3 GPM. Can you imagine what might happen if you install a pump that is rated at 5 GPM? If you guessed that the well might be pumped dry, you're right. The pump should not produce more water than the well can replenish. Keep this in mind when you are looking over the specifications provided to you by bidding subcontractors. Compare the pump rate with the well rate, and make sure that the pump is not too powerful for the well.

Pumps should not be suspended too close to the bottom of a well. How far from the bottom should a pump be placed? It depends on the static water level in the well. If a pump is placed too closely to the bottom of a well, it can pick up gravel, sand, and other debris. Some wells fill in a little over time, having a pump too close to the bottom when a well is filling in is bad news. A pump should be, in my opinion, at least 15 feet above the bottom, and higher when practical.

My personal pump is about 30 feet above the bottom of the well. The depth of my well is a little over 400 feet. The static level of my well water is only about 15 feet from the top of my well casing. This means that I have a column of water that is about 385 feet deep. That's a substantial reserve. I would have to pump hundreds of gallons of water at one time to run the well dry. Because I have so much water, it's easy for me to keep my pump hanging high above the bottom. Not all wells have so much reserve water, and this can force an installer to hang a pump closer to the bottom. The key is to keep the pump far enough from the bottom to avoid problems with debris being sucked into the pump.

SHALLOW WELLS

Dug wells and bored wells, or *shallow wells* as they are often called, are very different from drilled wells. While most drilled wells have diameters of 6 inches, a typical dug well will have a diameter of about 3 feet. Concrete usually surrounds a dug well, rather than the steel casing used for a drilled well. While drilled wells often reach depths of 300 feet, a dug well rarely runs deeper than 30 feet. There are, to be sure, many differences between the two types of wells.

In the old days, dug wells were created with picks, shovels, and buckets. It was dangerous work. Today, boring equipment is normally used to create a dug well. I suppose it would be more proper to call these wells bored wells, but most professionals I know refer to them as dug wells. To be specific on the type of well I'm talking about, let me explain the basic make-up of one.

If you see a concrete cylinder that has a diameter of approximately 3 feet sticking up out of the ground, you are looking at what I call a dug well. The concrete casing will typically be covered with a large, heavy concrete disk. You might want to call this type of well a bored well or a dug well, it's up to you. For my purposes, they are dug wells.

Old dug wells were dug by hand. Many of them were lined with stones. I've crawled down a few of these as a plumber, to work on pipes and such, but I don't think I'd do it today. My younger years as a plumber were more adventuresome than I would care to repeat.

Dug wells are always shallow in comparison with drilled wells. Finding a dug well that is 50 feet deep would be similar to finding pirate treasure sitting on top of a sandy beach. Most dug wells that I've seen have been no more than 35 feet deep. Bored wells can run much deeper. It's possible for a bored well to reach a depth of 100 feet, or more. The depth is regulated by the ground the well is dug in. Some types of earth allow for a deeper well than others. Still, in practical terms, most dug or bored wells won't exceed 50 feet in depth. Beyond this level, a drilled well is more practical.

Large-diameter wells go by many names. I call them dug wells. Sometimes they are bored wells. A lot of people know them as shallow wells. This is why jet pumps, which are often used with these types of wells, are called shallow-well pumps. Without trying to get extremely technical, I will simply refer to these wells as dug wells. You now know that I might be lumping bored wells into the category and that both types are considered shallow wells, so we shouldn't have a problem understanding each other.

Shallow wells are common in many parts of the country. When the water table is high and reasonably constant, shallow wells work fairly well. They might dry up during some hot, dry times of the year, but this is not always the case. Some shallow wells maintain a good volume of water all through the year.

Most of the homes that I've owned have had shallow wells. My present home has a drilled well, and I feel much more secure about it than I ever did with shallow wells. But, I only ever had trouble with one of my shallow wells, and that trouble only occurred for one summer. At other times and with my other wells, I never experienced any problems.

Shallow wells are much less expensive to install than drilled wells are. This is one good reason for using a shallow well. However, there are drawbacks to a shallow well. Having a sufficient water quantity is one of these drawbacks. Due to the large diameter of a dug well, a lot of water can be stored in reserve. While the water in a shallow well might be only a few feet deep, it has a lot more surface area than water stored in a deep well. The increased surface area helps to make up for the lack of depth.

Even with a large diameter, dug wells do often run dry for short periods of time. If they don't run completely dry, they might contain such a small quantity of water that rationing is needed. For example, you might have to go to a local laundromat to wash clothes for a few weeks out of the year. This, of course, is more inconvenience than some homeowners are willing to put up with.

When Should You Use a Shallow Well?

When should you use a shallow well? The ground conditions where the well will be drilled have a lot of influence on this decision. If bedrock is near the ground surface, a dug well is not practical. Drilled wells are the answer when you have to penetrate hard rock. To give you a good idea of what depths are possible and in what types of soil conditions different types of wells can be used, let me break them down into categories.

True Dug Wells. True dug wells can be created with depths of up to 50 feet. Their diameter can range from a common 3 feet to a massive 20 feet. In terms of geographic formations, clay, silt, sand, gravel, cemented gravel, and even boulders can be dealt with. Sandstone and limestone might allow the use of a dug well, but the material must either be soft or fractured. Dense igneous rock cannot be penetrated with a dug well.

Bored Wells. Bored wells can reach depths of 100 feet. The diameter of this type of well can be a minuscule 2 inches. Larger diameters, in the 30-inch range are also common. Clay, silt, sand, and gravel can all support a bored well. Cemented gravel will stop a bored well, and so will bedrock. Boulders can sometimes be worked around, and sandstone and limestone affect a bored well in the way that they do a dug well.

Drilled Wells. Drilled wells can run to depths of 1000 feet. This is far from being a shallow well. All of the geologic formations mentioned for the previous well types can be overcome with a drilled well. Percussion and hydraulic rotary drilling are very similar in their abilities.

Jetted Wells. Jetted wells can run up to 100 feet in depth. They cannot be installed in bedrock, limestone, or sandstone. Boulders, cemented gravel, and large loose gravel all prohibit the use of jetted wells. Clay, sand, and silt are the best types of geologic formations to use a jetted well in. The diameter of a jetted well can be anywhere from 2 to 12 inches.

Characteristics

Let's talk about some of the characteristics of shallow wells. Dug wells can extend only a few feet below a water table. The geologic conditions affect the recovery rate of a well. For example, a dug well that is surrounded by gravel might produce an excellent recovery rate, while one having fine sand surrounding it might produce a poor rate of recovery. Dug wells cannot be considered terrific water producers.

Bored wells are a little different. They can extend deeper into water table. Going 10 feet below the edge of a water table is not uncommon. This added depth gives bored wells an advantage over dug wells. Having an 8-foot head of water with a 3-foot diameter provides a good supply of water, assuming that the recovery rate is good. It would not be considered strange or unusual for a bored well to have a recovery rate of 20 GPM. This is about double what would be expected of a good dug well.

Even though many professionals, myself included, call all shallow wells dug wells, there is clearly a difference between a bored well and a true dug well. Most shallow wells installed today are bored wells. If your well installer talks about a dug well, make sure that the installer is actually referring to a bored well.

Some bored wells hit artesian aquifers. When this happens, the static water level in the well rises. For example, a bored well that extends 10 feet past the water table would normally be thought of as holding 10 feet of water. If an artesian aquifer is hit, the actual water reserve could be 20 feet deep, or more. I've seen bored wells with such powerful artesian effects that water actually ran out over the top of the well casing. It is certainly possible to obtain plenty of water for a residence with a bored well. However, there is no guarantee of it.

The Preliminary Work

The preliminary work for a builder who is having a shallow well installed is very similar to that of a builder who is preparing for the installation of a drilled well. Because we covered the steps for this in the last chapter, I won't repeat them here. The only big difference is the size of a shallow well. A drilled well can be hidden easily with a few strategically placed shrubs. The casing and cap of a shallow well is much harder to hide. Aside from the size of the well, the rest of the steps are about the same.

AFTER A WELL IS IN

After a well is in, you are faced with the disinfection process. Just as we discussed in the last chapter, chlorine is normally used to purify a well before testing is done on the water. There are a few extra things to consider, however, when testing a shallow well. For one thing, the well cover is heavy. One person can slide the cover aside, but it is easier with two people. The concrete cover is brittle. Rough handling can cause it to crack or break. Having a concrete cover drop and catch your finger between the lip of the well casing and the cover could be extremely painful. Be careful when handling the cover.

The odds of a child or an animal falling into an uncovered drilled well are much lower than they are with a shallow well. Two adults could jump into the opening of a bored well without any problem. Never leave the cover off of a shallow well when you are not right at the well site. Even if you are just going into the house to turn the water on or off, put the cover back on the well. Pets and people can fall into an open well quickly, and the results can be disastrous. Even drilled wells should be covered every time they are left unattended. Curious kids and pets can find themselves in some very dangerous circumstances in the blink of an eye.

When you are purging chlorine from a shallow well, you must monitor the water level closely. While it is uncommon for a drilled well to be pumped dry during a purging, it is not unusual for a dug or bored well to have it's water level fall below the foot valve or end of the drop pipe. If this happens, the pump keeps pumping, but it's only getting air. This is bad for the pump and can burn it up. Monitor the water level closely as you empty a shallow well.

You can watch the water level in most shallow wells without any special equipment. Your eyes should be the only tools needed to tell when the water level drops down too far. However, a string with a weight attached to it can be used to gauge the water depth. If you know how long the drop pipe in the well is (your pump installer should be able to tell you), make the string a foot shorter than the drop pipe. As long as the string hits water each time it is dropped into the well, you know the water level is safe.

PUMP SELECTION

Pump selection for a shallow well is sometimes a little more complex than it is for a deep well. Almost everyone uses submersible pumps for drilled wells. It's possible to use a two-pipe jet pump for a deep well, but few installers recommend this course of action. Shallow wells, on the other hand, can use single-pipe jet pumps, two-pipe jet pumps, or even submersible pumps. The use of submersible pumps in shallow wells is not common. In wells where the water lift is not more than about 25 feet, single-pipe jet pumps are often used. Deeper bored wells, and some dug wells, see the use of two-pipe jet pumps. If a well of any type has an adequate depth of water to work with, a submersible pump can always be used.

Single-pipe Jet Pumps

Single-pipe jet pumps are the least expensive option available for a shallow well. However, their lifting capacity is limited. Because single-pipe jet pumps work on a suction-only basis, they must be able to pull a vacuum on the water. Because physics plays a part in how high above sea level a vacuum can be made, jet pumps can only pull water so high. Without looking it up, I can't remember the exact maximum lift under ideal conditions. I know it's around 30 feet. However, for practical purposes, most professionals agree that a suction-based pump should not be expected to lift water more than 25 feet.

Jet pumps are installed at some location outside of the well. Unlike submersible pumps, which hang in the well water, jet pumps are normally installed in basements, crawl spaces, or pump houses. The pumps and their pipes must be protected from freezing temperatures. A pressure tank should be used with jet pumps, just as they are with submersible pumps. Many jet pumps are made to sit right on top of a pressure tank, with the use of a special bracket. The pressure tank must also be protected from freezing temperatures.

Two-pipe Jet Pumps

Two-pipe jet pumps can be used to pump water from deeper wells. These pumps use two pipes. The pump forces water down one pipe to allow it to be sucked up in the other. Because a two-pipe pump is not dependent solely on suction, it can handle higher lifts. The physical appearance of a two-pipe pump is basically the same as that of a one-pipe pump, except for the extra pipe.

Submersible Pumps

Submersible pumps push water up out of a well. Because the pump is pushing, rather than sucking, it can be installed at great depths. In many ways, submersible pumps are far less prone to failure than jet pumps. They are more expensive, but they tend to last longer, and there are not as many parts and pieces to fail. This makes submersible pumps a favorable choice when the water depth is sufficient to warrant their use.

Because jet pumps are most often used with shallow wells, we will concentrate our efforts on them. Submersible systems were described in the last chapter, if you want to review them. Because most shallow wells are equipped with a single-pipe jet pump, we will start our installation procedures with them.

INSTALLING A SINGLE-PIPE JET PUMP

Installing a single-pipe jet pump is fairly easy. The well portion of the work is particularly simple. Almost anyone with modest mechanical skills and an ability to read instructions can manage the installation of a jet pump. However, a license to perform this type of work is probably required in your area. This might prohibit you from making your own installations.

The first step towards installing a jet pump will be digging a trench between the well casing and the pump location. Because we covered trenching in the last chapter, we will skirt the issue here. Once the trench is dug, you are ready to make a hole in the side of the well casing. Because most shallow wells are cased with concrete, a cold chisel and a heavy hammer is all that's needed to make a hole. The hole will have to be patched to make it watertight after the well piping is installed. Otherwise, ground water will run into the well and might contaminate it.

After a hole is made, you are ready to see the well pipe installed. The pipe will probably be PE pipe. Insert fittings should be made of metal, in my opinion, and all connections should be double clamped. A foot valve is usually installed on the end of the pipe that will hang in the well water. The foot valve serves two purposes. It acts as a strainer to block out gravel and similar debris that might otherwise be sucked into the pipe and pump. Foot valves also act as check valves, to prevent water in the drop pipe from running out into the well. If the water in a drop pipe was not controlled with a check valve or foot valve, the jet pump would lose its prime and fail to pump water.

After the foot valve is installed, an elbow fitting is attached to the other end of the pipe. This fitting will allow the water service pipe, which will be placed in the trench, to connect to the drop pipe. Some form of protection should be provided for the water service pipe where it penetrates the concrete casing. Form pipe insulation can sometimes be used, and rigid plastic pipe, like the type used for drains and vents in plumbing systems, can always be used. If PE pipe is allowed to rest on the rough edge of concrete, constant vibration of the pipe when the pump is pumping will eventually wear a hole through the pipe.

After the connection is made between the drop pipe in the well and the water service pipe in the trench, the rest of the work will be done at the pump location. Backfilling the trench should be done with the same precautions described in the last chapter.

Jet pumps can be installed almost anywhere where they will not get wet or become frozen. Crawl spaces, basements, pump houses, and even closets are all potential installation locations. A pressure tank is usually installed in close proximity to a jet pump. In some cases, the pump will be bolted to a bracket on a pressure tank. Sometimes a small pressure tank is suspended above a pump. Many tanks stand independently on a solid surface, such as a floor. Part of the decision of the type of tank used and its location will be based on the space available for the tank and pump. It is not mandatory that a pressure tank be used, but it is highly recommended, to prolong the life of a pump. However, a pressure tank does not have to be installed adjacent to a pump. It can be in a remote location.

The piping arrangement for a jet pump is not complicated. There are, however, many accessories that are commonly used, such as an air-volume control. Shut-off valves are, of course, installed in the pipes leaving the pump. In the case of a one-pipe jet pump, one large pipe brings water to the pump. A smaller pipe distributes the water to a water distribution system. A pressure switch is needed to control when the pump cuts on and off. Relief valves are needed to protect pressure tanks. Drain valves are typically installed for pressure tanks, and a check valve might be installed.

TWO-PIPE PUMPS

Two-pipe pumps look very similar to one-pipe pumps, except for the extra pipe involved. While one-pipe jet pumps have only a suction pipe, two-pipe pumps have both a pressure and a suction pipe. There are some differences in the piping arrangement in the well to go along with this most noticeable difference. An ejector is installed to enable the pressure pipe to assist the suction pipe in lifting water from the well.

A pressure tank should still be used with a two-pipe system. In fact, a pressure tank should be installed with all types of well pumps for residential use. We've talked about pressure tanks in terms of their existence, but let's spend some time talking about them in detail.

PRESSURE TANKS

Pressure tanks should be considered standard equipment with every residential pump installation. The tanks are not very expensive, and they can add years to the life of a pump. They can also provide residents with better water pressure, which is an important aspect to consider. It's very rare today to see a plumbing system that is fed by a well where a pressure tank is not used. I can't imagine a pump installer bidding a job without including the cost of a pressure tank, but make sure that your next well system does provide a pressure tank. With that said, let's talk about the specifics of various sizes and types of pressure tanks.

Small Tanks

Small tanks are available in sizes that hold no more than 2 gallons of water. This is a very small tank. In my opinion, it is too small for almost any application. A tank that has such a minimum capacity will do almost no good in a routine, residential plumbing system. When you consider that many toilets use more than 2 gallons of water each time they are flushed, you can see that the pump will be running a lot. Think about showers. At a flow rate of 3 gallons per minute, the first minute of a shower will deplete the supply of a super small pressure tank.

The whole purpose of installing a pressure tank is to take some strain off of a pump. While a 2-gallon buffer provides some support to a pump, the help is minimal. A pressure tank should be sized to meet the needs of the house it serves. In other words, the number of people using the plumbing system and the types of fixtures available should be taken into consideration when choosing a pressure tank.

Large Tanks

Large tanks can take up a substantial amount of room. This is not normally a problem in houses where a basement, cellar, or crawl space is available. However, if the pump system and tank must be installed within the primary living space of a home, tank size can become an issue. Residential pressure tanks can be purchased with an ability to hold more than 100 gallons of water. In most circumstances, this is overkill. A tank that holds between 20 and 40 gallons should perform well under average residential conditions.

In-line Tanks

In-line tanks are designed to be installed right off of a water pipe. They can be suspended below the pipe, or they can rise above the pipe. The size of in-line tanks vary. A brand that I use offers in-line tanks with capacities of 2 gallons, about 4½ gallons, about 8½ gallons, a little over 10 gallons, and 14 gallons. This range of sizes can be adequate for small homes.

An in-line tank is not normally mounted on a bracket or sat on a floor. It typically hangs from a pipe or sticks up above the pipe. When floor space is at a premium, an in-line tank is desirable. As a builder, I would try to keep the size up in the 10-gallon range as a minimum. However, I recently installed a smaller tank for a summer cottage. It was one of the 4½-gallon models. Because the cottage had only one bathroom and rarely accommodated more than two people, the small tank was deemed adequate. You must match your tank to your needs.

Stand Models

Stand models are the type of pressure tank most often installed in homes. These units sit on a floor in a free-standing mode. Piping is run from the pump to the tank and then from the tank to the water distribution system. With capacities ranging from about 10 gallons up to 119 gallons, this type of tank can meet any residential need.

I have a stand model in my home. It's capacity is around 40 gallons. This tank cost more to buy than a smaller tank would, but its size allows my pump to run less often. This saves electricity and prolongs the life of my pump. I think the trade-off is worthwhile.

When a stand model is used, it is easiest to install when a tank tee, or a *tank cross* as they are often called, is used. This is a fitting that is designed for use with a pressure tank. The tank tee screws into the pressure tank and provides both an inlet and outlet opening. In addition to these openings, the fitting is tapped for accessories, such as pressure gauges and relief valves. The use of this fitting not only makes a job easier to install, it gives the appearance of a more professional installation.

Pump-stand Models

Pressure tanks are available in pump-stand models. These tanks are frequently used in conjunction with jet pumps. They are not needed with submersible pumps, because submersibles are hung in a well. A pump-stand model is designed to sit horizontally. It has a bracket on top of it so that a pump can be bolted down to it.

Using a pump-stand model is one way to conserve floor space. Because a pump attaches to the top of the stand, there is only one object sitting on the floor. With a typical stand model, both the pressure tank and the pump would be installed on the floor. From the manufacturer who I deal with, there are two sizes of pump-stand tanks available. One is about 8½ gallons, and the other is 14 gallons. I would opt for the latter.

Underground Tanks

Although I have never installed one, there are underground tanks available. Their size ranges from 14 gallons to 62 gallons when purchased from the manufacturer of my choice. Personally, I can't think of a time when an underground tank would have helped me, but I'm sure there must be occasions when they are desirable.

Diaphragm Tanks

Diaphragm pressure tanks are common in today's plumbing systems. This was not always the case. Older houses often have plain old galvanized storage tanks. These standard tanks can be a real pain to live with. They frequently become waterlogged. By this, I mean that they lose their air content and fill with water. When this happens, the tank must be drained, pumped up with air, and refilled with water. A waterlogged tank will make a pump run every time water is called for at a faucet, thus eliminating the advantage of having a pressure tank.

Modern pressure tanks come precharged with air, and they have a diaphragm system that eliminates waterlogging. Standard tanks are still available, but they are rarely used. I suggest that you specify a diaphragm tank in your well specs.

Another problem with a standard tank is rust. After some period of time, a metal holding tank will begin to rust. Air will leak out, and eventually, water will leak. This means a patch or replacement will have to be made. With modern pressure tanks, liners are used so that water never comes into contact with the metal housing of the tank. This, of course, eliminates the possibility of interior rust. Bacteria growth and rust in drinking water are not nearly as likely with a lined tank as they are with a traditional metal tank.

Sizing

Sizing a pressure tank is an important step in designing and installing a good pump system. The pressure tank protects the pump. I mentioned earlier that you should take the number of people and plumbing fixtures into consideration when thinking of a size for your pressure tank. My comment was meant to get you thinking about how little use it takes to drain down a small tank. If you want to seriously size a tank, you should work from the specifications for the pump that will be installed. In other words, the ideal sizing comes from carefully matching your tank to your pump. The relevance of a number of people or plumbing fixtures plays only a small role, if any, in determining tank size.

Pressure tanks are designed to work with pressure switches. These switches can be set for various cut-in and cut-out pressures. It has long been common for a pump to cut on when tank pressure drops to 20 pounds per square inch (PSI). At this cut-in rate, a typical cut-out rate is 40 PSI. For years, this has been something of a standard in well systems. However, times change, and so do standards.

It is not at all uncommon for cut-in pressures to be set at 30 PSI today. A corresponding cut-out pressure is 50 PSI. Some homes have their well systems set up with cut-in pressures of 40 PSI and cut-out pressures of 60 PSI. The increase in pressure is due to many factors. People tend to enjoy increased water pressure at some of their plumbing fixtures, such as showers. Some plumbing fixtures and devices require higher working pressures than what was common in the past. This all contributes to the trend for higher pressures.

Before you can size a pressure tank, you must determine what your cut-in and cut-out pressures will be. You must also know the gallons-per-minute (GPM) rating of the pump. A time period must be established as to the minimum run time for the pump. This is usually established by the manufacturer, so check your pump paperwork to see what the minimum run time should be.

Once you know all of the previously mentioned variables, you can select a tank of the proper size. However, you need some type of sizing chart to do this math. Most manufacturers of pressure tanks will be happy to provide you with sizing tables.

Installation Procedures

Installation procedures for pressure tanks can take many forms. You've already seen that there are numerous types of tanks available. The installation of these tanks vary as much as the tanks do. You might install an underground tank to work in conjunction with a submersible pump. A pump-stand tank is sometimes a good choice for a jet pump. In-line tanks can also come in handy. Stand models are the type that you will use most often. While it is highly unlikely in a residential system, unless you are building a house on a farm, a multiple-tank setup might be called for. Your installer certainly should know how to make a good installation.

A Relief Valve. Make sure a relief valve is installed to protect the pressure tank from excessive pressure. Omitting this fitting could result in disaster. It is a code requirement, so someone should catch it if you and your installer miss it, but don't take this chance. Ratings on pressure tanks and relief valves can vary, but most tanks are rated at 100 PSI, and the relief valves used with them are rated at 75 PSI. If a relief valve isn't installed, a pressure tank can blow up, causing personal and property damage. The same thing can happen if the relief valve is rated higher than the tank's working pressure. This is a serious issue, so take an active part in making sure the right relief valve is installed properly.

Keep It Dry. A pressure tank will last longer when you keep the exterior of it dry. In some cases, this might require installing a tank on blocks or a platform. If the basement, cellar, or other area where a tank is being installed tends to get wet during some seasons, elevate the tank to protect it. This is very simple to do at the time of installation, but it becomes a bit complicated after the fact. Spending a few bucks for some blocks can save your customers money down the road. Happy customers are what successful businesses are all about, so don't forget them when you are putting out the specs on a job.

CHAPTER 4
BASIC WELL PROBLEMS

Basic well problems are not common. In this category, we are talking about trouble with wells themselves, more so than with pumps. Water quality is not a part of our present discussion. Because well problems are often segregated by the type of well being used, we will investigate the problems on the basis of specific wells.

DRIVEN WELLS

Driven wells don't usually have a large holding capacity. It is not uncommon for flow rates or recovery rates for driven wells to be low. Both of these factors can contribute to a person running out of water. If a customer calls with a complaint of having no water when a driven well is in use, you might be faced with a pump problem or a well problem. This is true of all types of wells. Driven wells are the most likely type of well to be run dry. This is a simple problem to troubleshoot.

If you suspect that a driven well is out of water, you can gain access to the well and drop a weighted line into the well to determine if any water is standing in the well. If there is little or no water present, your problem might be with the point filter or the water source. If the well point has become clogged, water will not be able to enter the well pipe. Assuming that there is insufficient water, you can take one of two actions.

You could pull the well point and inspect it. This will not be an easy task. If your area is experiencing an extremely dry spell, you might have to pull the well point to identify the problem. However, if area conditions don't point to a drop in local water tables, you might ask the customer to avoid using any water for a few hours and then try the pump again. Given several hours for recovery, the well might produce a new supply of water. This doesn't rule out a partially clogged point, but it tends to indicate a low flow rate. To be sure of what is going on, the point will have to be pulled and inspected.

Due to the nature of a driven point, your options for finding out if water is present in the water table are limited. If water can't pass through the filter of a point, water won't enter the well. This is not the case with other types of wells. Unfortunately, some of the money saved by installing a well point can be lost through later problems, such as having to pull the point for inspection and possible replacement.

Sand

Sand in a water distribution system that is served by a well point is an indication that the screen filter on the well point used has openings that are too large. This type of problem can be addressed with the addition of an in-line sediment filter, but the true solution lies in replacing the well point with one where a finer screen filter is used.

Other Contaminants

Other contaminants can enter a water distribution system through a well point. These entries into the water system can be filtered out with water treatment condition. Replacing a well point might help to solve this type of problem. Essentially, some type of conditioning equipment will probably be needed to eliminate very small contaminants.

SHALLOW WELLS

Shallow wells are less likely to pose problems than driven wells are. However, these wells can provide builders with head-scratching trouble. Some shallow wells cave in over time. It doesn't always take a long time for this to happen. A new well can experience problems with cave-ins long before warranty periods are over. This is not a common problem, but it is one that could exist.

Shallow wells do sometimes run out of water. Given some time, these wells normally recover a water supply. If a shallow well runs dry, there is very little that can be done. You must either wait for water to return or create a new well.

Problems sometime exist where sand or other sediment is pumped out of a shallow well. This is usually a result of having a foot valve or drop pipe hung too low in the well. If a well has worked well for a few months and then begins producing sand or other particles, it can be an indication that the well is caving in. Sometimes a foot valve will become clogged under these conditions. The simple act of shaking the drop pipe can clear a foot valve of debris and allow a pump to return to normal operation.

Checking a shallow well to see if water is in reserve is easy. A weighted line can be dropped into a well to establish water depth. Assuming that water is present in sufficient quantity, you can rule out a dry well. However, you cannot rule out the fact that the drop pipe or foot valve in the pump might be installed above the water level. This can be checked by pulling the drop pipe out of the well and measuring it. The length of a drop pipe can then be compared to the depth at which water is located. If you have water in the well and your drop pipe is submerged in it, you can rule the well out as your problem.

DRILLED WELL

Drilled wells very rarely run out of water. It is, however, possible that a drilled well could be run dry. A weighted line will allow you to test for existing water. Pulling the drop pipe and comparing its length to the depth at which water is contacted will prove if a problem of not having water in a house is the well's fault or some fault of a pumping system.

In all of my years as a plumber, I've never known a drilled well to run out of water without some type of outside interference. By outside interference, I mean some form of man-made trouble, such as blasting with explosives somewhere in the general area. Let me give you an example of this from my recent past.

A friend of mine has enjoyed a drilled well for decades. In all of these years, the well had never given its owner any problem until this past summer. Road work was being done within a mile or so of my friends house in the early summer. Part of the work involved the blasting of bedrock. Shortly after this blasting took place, my friends well quit producing water. Why? My guess (it's only a guess) is that the blasting caused a change in the underground water path. It might be that the blasting shifted the rock formations and diverted the water that was at one time serving the well of my friend. I've seen similar situations occur at other times. It's impossible for me to say with certainty that blasting ruined the well, but it's my opinion that it did.

TROUBLESHOOTING JET PUMPS

You are already aware that there are differences between jet pumps and submersible pumps. Knowing this, it only makes sense that there will be differences in the types of problems encountered with the different types of pumps. Let's start our troubleshooting session with jet pumps.

Will Not Run

A pump that will not run can be suffering from one of many failures. The first thing to check is the fuse or circuit breaker to the circuit. If the fuse is blown, replace it. When the circuit breaker has tripped, reset it. This is something you could ask your customer to check.

When the fuse or circuit breaker is not at fault, check for broken or loose wiring connections. Bad connections account for a lot of pump failures. It is possible the pump won't run due to a motor overload protection device. If the protection contacts are open, the pump will not function. This is usually a temporary condition that corrects itself.

If the pump is attempting to operate at the wrong voltage, it might not run. Test the voltage with a voltammeter. The power must be on when this test is conducted. With the leads attached to the meter and the meter set in the proper voltage range, touch the black lead to the white wire and the red lead to the black wire in the disconnect box near the pump. Test both the incoming and outgoing wiring.

Your next step in the testing process should be at the pressure switch. The black lead should be placed on the black wire, and the red lead should be put on the white wire for this test. There should be a plate on the pump that identifies the proper working voltage. Your test should reveal voltage that is within 10% of the recommended rating.

An additional problem that you might encounter is a pump that is mechanically bound. You can check this by removing the end cap and turning the motor shaft by hand. It should rotate freely.

A bad pressure switch can cause a pump to not run. With the cover removed from the pressure switch, you will see two springs, one tall and one short. These springs are depressed and held in place by individual nuts. The short spring is preset at the factory and should not need adjustment. This adjustment controls the cut-out sequence for the pump. If you turn the nut down, the cut-out pressure will be increased. Loosening the nut will lower the cut-out pressure.

The long spring can be adjusted to change the cut-in and cut-out pressure for the pump. If you want to set a higher cut-in pressure, turn the nut tighter to depress the spring further. To reduce the cut-in pressure you should loosen the nut to allow more height on the spring. If the pressure switch fails to respond to the adjustments, it should be replaced.

It is also possible that the tubing or fittings on the pressure switch are plugged. Take the tubing and fittings apart and inspect them. Remove any obstructions, and reinstall the tubing and fittings.

The last possibility for the pump failure is a bad motor. You will use an Ohmmeter to check the motor, and the power to the pump should be turned off. Start checking the motor by disconnecting the motor leads. We will call these leads L1 and L2. The instructions you are about to receive are for Goulds pumps with motors rated at 230 volts. When you are conducting the test on different types of pumps, you should refer to the manufacturer's recommendations.

Set the Ohmmeter to RX100, and adjust the meter to zero. Put one of the meter's leads on a ground screw. The other lead should systematically be touched to all terminals on the terminal board, switch, capacitor, and protector. If the needle on your Ohmmeter doesn't move as these tests are made, the ground check of the motor is okay.

The next check to be conducted is for winding continuity. Set the Ohmmeter to RX1, and adjust it to zero. You will need a thick piece of paper for this test; it should be placed between the motor switch points, and the discharge capacitor.

You should read the resistance between L1 and A to see that it is the same as the resistance between A and yellow. The reading between yellow to red should be the same as L1 to the same red terminal.

The next test is for the contact points of the switch. Set the Ohmmeter to RX1, and adjust it to zero. Remove the leads from the switch, and attach the meter leads to each side of the switch; you should see a reading of zero. If you flip the governor weight to the run position, the reading on your meter should be infinity.

Now let's check the overload protector. Set your meter to RX1, and adjust it to zero. With the overload leads disconnected, check the resistance between terminals one and two and then between two and three. If a reading of more than one occurs, replace the overload protector.

The capacitor can also be tested with an Ohmmeter. Set the meter to RX1000, and adjust it to zero. With the leads disconnected from the capacitor, attach the meter leads to each terminal. When you do this, you should see the meter's needle go to the right and drift slowly to the left. To confirm your reading, switch positions with the meter leads and see if you get the same results. A reading that moves toward zero or a needle that doesn't move at all indicates a bad capacitor.

I realize the instructions I've just given you might seem quite complicated. In a way, they are. Pump work can be very complex. I recommend that you leave major troubleshooting to the person who installed your problem pump. If you are not familiar with controls, electrical meters, and working around electrical wires, you should not attempt many of the procedures I am describing. The depth of knowledge I'm providing you might be deeper than you ever expect to use, but it will be here for you if you need it.

Runs but Gives No Water

When a pump runs but gives no water, you have seven possible problems to check out. Let's take a look at each troubleshooting phase in their logical order.

The first consideration should be that of the pump's prime. If the pump or the pump's pipes are not completely primed, water will not be delivered. For a shallow-well pump, you should remove the priming plug and fill the pump completely with water. You might want to disconnect the well pipe at the pump and make sure it is holding water. You could spend considerable time pouring water into a priming hole only to find out the pipe was not holding the water.

For deep-well jet pumps, you must check the pressure-control valves. The setting must match the horse power and jet assembly used, so refer to the manufacturer's recommendations.

Turning the adjustment screw to the left will reduce pressure, and turning it to the right will increase pressure. When the pressure-control valve is set too high, the air-volume control cannot work. If the pressure setting is too low, the pump might shut itself off.

If the foot valve or the end of the suction pipe has become obstructed or is suspended above the water level, the pump cannot produce water. Sometimes shaking the suction pipe will clear the foot valve and get the pump back into normal operation. If you are working with a two-pipe system, you will have to pull the pipes and do a visual inspection. However, if the pump you are working on is a one-pipe pump, you can use a vacuum gauge to determine if the suction pipe is blocked.

If you install a vacuum gauge in the shallow-well adapter on the pump, you can take a suction reading. When the pump is running, the gauge will not register any vacuum if the end of the pipe is not below the water level or if there is a leak in the suction pipe.

An extremely high vacuum reading, such as 22 inches or more, indicates that the end of the pipe or the foot valve is blocked or buried in mud. It can also indicate the suction lift exceeds the capabilities of the pump.

A common problem with the symptoms of the pump running without delivering water is a leak on the suction side of the pump. You can pressurize the system and inspect it for these leaks.

The air-volume control can be at fault for a pump that runs dry. If you disconnect the tubing and plug the hole in the pump, you can tell if the air-volume control has a punctured diaphragm. If plugging the pump corrects the problem, you must replace the air-volume control.

Sometimes the jet assembly will become plugged up. When this happens with a shallow-well pump, you can insert a wire through the ½" plug in the shallow-well adapter to clear the obstruction. With a deep-well jet pump, you must pull the piping out of the well and clean the jet assembly.

An incorrect nozzle or diffuser combination can result in a pump that runs but that produces no water. Check the ratings in the manufacturer's literature to be sure the existing equipment is the proper equipment.

The foot valve or an in-line check valve could be stuck in the closed position. This type of situation requires a physical inspection and the probable replacement of the faulty part.

Cycles Too Often

When a pump cycles on and off too often, it can wear itself out prematurely. This type of problem can be caused by several types of problems. For example, any leaks in the piping or pressure tank would cause frequent cycling of the pump.

The pressure switch might be responsible for a pump that cuts on and off to often. If the cut-in setting on the pressure gauge is set too high, the pump will work harder than it should.

If the pressure tank becomes waterlogged (filled with too much water and not enough air), the pump will cycle frequently. If the tank is waterlogged, it will have to be recharged with air. This would also lead you to suspect the air-volume control of being defective.

An insufficient vacuum could cause the pump to run too often. If the vacuum does not hold at 3" for 15 seconds, it might be the problem.

The last thing to consider is the suction lift. It's possible that the pump is getting too much water and creating a flooded suction. This can be remedied by installing and partially closing a valve in the suction pipe.

Won't Develop Pressure

Sometimes a pump will produce water but will not build the desired pressure in the holding tank. Leaks in the piping or pressure tank can cause this condition to occur.

If the jet or the screen on the foot valve is partially obstructed, the same problem might result.

A defective air-volume control might prevent the pump from building suitable pressure. You can test for this by removing the air-volume control and plugging the hole where it was removed. If this solves the problem, you know the air-volume control is bad.

A worn impeller hub or guide vane bore could result in a pump that will not build enough pressure. The proper clearance should be .012 on a side or .025, diametrically.

With a shallow-well system, the problem could be caused by the suction lift being too high. You can test for this with a vacuum gauge. The vacuum should not exceed 22 inches at sea level. Deep-well jet pumps require you to check the rating tables to establish their maximum jet depth. Also with deep-well jet pumps, you should check the pressure-control valve to see that it is set properly.

Switch Fails

If the pressure switch fails to cut out when the pump has developed sufficient pressure, you should check the settings on the pressure switch. Adjust the nut on the short spring, and see if the switch responds; if it doesn't, replace the switch.

Another cause for this type of problem could be debris in the tubing or fittings between the switch and pump. Disconnect the tubing and fittings, and inspect them for obstructions.

TROUBLESHOOTING SUBMERSIBLE PUMPS

There are some major differences between troubleshooting submersible pumps and troubleshooting jet pumps. One of the most obvious differences is that jet pumps are installed outside of wells and submersible pumps are installed below the water level of wells.

There are times when a submersible pump must be pulled out of a well, and this can be quite a chore. Even with today's lightweight well pipe, the strength and endurance needed to pull a submersible pump up from a deep well is considerable. Plumbers that work with submersible pumps regularly often have a pump-puller to make removing the pumps easier.

When a submersible pump is pulled, you must allow for the length of the well pipe when planning the direction to pull from and where the pipe and pump will lay once removed from the well. It is not unusual to have between 100 and 200 feet of well pipe to deal with, and some wells are even deeper.

It is important when pulling a pump, or lowering one back into a well, that the electrical wiring does not rub against the well casing. If the insulation on the wiring is cut, the pump will not work properly. Let's look now at some specific troubleshooting situations.

Won't Start

A pump that won't start might be the victim of a blown fuse or tripped circuit breaker. If these conditions check out okay, turn your attention to the voltage.

In the following scenarios, we will be dealing with Goulds pumps and Q-D-type control boxes.

To check the voltage, remove the cover of the control box to break all motor connections. Be advised: Wires L1 and L2 are still connected to electrical power. These are the wires running to the control box from the power source.

Press the red lead from your voltmeter to the white wire and the black lead to the black wire. Keep in mind that any major electrical appliance that might be running at the same time the pump would be, like a clothes dryer, should be turned on while you are conducting your voltage test.

Once you have a voltage reading, compare it to the manufacturer's recommended ratings. For example, with a Goulds pump that is rated for 115 volts, the measured volts should range from 105 to 125. A pump with a rating of 208 volts should measure a range from 188 to 228 volts. A pump rated at 230 volts should measure between 210 and 250 volts.

If the voltage checks out okay, check the points on the pressure switch. If the switch is defective, replace it.

The third likely cause of this condition is a loose electrical connection in either the control box, the cable, or the motor. Troubleshooting for this condition requires extensive work with your meters.

To begin the electrical troubleshooting, we will look for electrical shorts by measuring the insulation resistance. You will use an Ohmmeter for this test, and the power to the wires you are testing should be turned off.

Set the Ohmmeter scale to RX100K, and adjust it to zero. You will be testing the wires coming out of the well, from the pump, at the well head. Put one of the Ohmmeter's leads to any one of the pump wires, and place the other Ohmmeter lead on the well casing or a metal pipe. As you test the wires for resistance, you will need to know what the various readings mean, so let's examine this issue.

You will be dealing with normal Ohm values and megohm values. Insulation resistance will not vary with ratings. Regardless of the motor, horsepower, voltage, or phase rating, the insulation resistance will remain the same.

A new motor that has not been installed should have an Ohm value of 20,000,000 or more and a megohm value of 20. A motor that has been used but is capable of being reinstalled should produce an Ohm reading of 10,000,000 or more and a megohm reading of 10.

Once a motor is installed in the well, which will be the case in most troubleshooting, the readings will be different. A new motor installed with its drop cable should give an Ohm reading of 2,000,000 or more and a megohm value of 2.

An installed motor in a well that is in good condition will present an Ohm reading of between 500,000 and 2,000,000. Its megohm value will be between 0.5 and 2.

A motor that gives a reading in Ohms of between 20,000 and 500,000 and a megohm reading of between 0.02 and 0.5 might have damaged leads or might have been hit by lightning; however, don't pull the pump yet.

You should pull the pump when the Ohm reading ranges from 10,000 to 20,000 and the megohm value drops to between 0.01 and 0.02. These readings indicate a motor that is damaged or cables that are damaged. While a motor in this condition might run, it probably won't run for long.

When a motor has failed completely or the insulation on the cables has been destroyed, the Ohm reading will be less than 10,000 and the megohm value will be between 0 and 0.01.

With this phase of the electrical troubleshooting done, we are ready to check the winding resistance. You will have to refer to charts as reference for correct resistance values, and you will have to make adjustments if you are reading the resistance through the drop cables. I'll explain more about this in a moment.

If the Ohm value is normal during your test, the motor windings are not grounded and the cable insulation is intact. When the Ohm readings are below normal, you will have discovered that either the insulation on the cables is damaged or the motor windings are grounded.

To measure winding resistance with the pump still installed in the well, you will have to allow for the size and length of the drop cable. Assuming you are working with copper wire, you can use the following figures in Figure 4.1

Cable size	Resistance
14	.5150
12	.3238
10	.2036
8	.1281
6	.08056
4	.0506
2	.0318

FIGURE 4.1 The resistance of cable for each 100 feet in length and ohms per pair of leads.

to obtain the resistance of cable for each 100 feet in length and Ohms per pair of leads.

If aluminum wire is being tested, the readings will be higher. Divide the Ohm readings above by 0.61 to determine the actual resistance of aluminum wiring.

If you pull the pump and check the resistance for the motor only (not with the drop cables being tested), you will use different ratings. You should refer to a chart supplied by the manufacturer of the motor for the proper ratings.

When all the Ohm readings are normal, the motor windings are fine. If any of the Ohm values are below normal, the motor is shorted. An Ohm value that is higher than normal indicates that the winding or cable is open or that there is a poor cable joint or connection. Should you encounter some Ohm values being higher than normal while others are lower than normal, you have found a situation where the motor leads are mixed up and need to be attached in their proper order.

If you want to check an electrical cable or a cable splice, you will need to disconnect the cable and have a container of water to submerge the cable in, a bathtub will work.

Start by submerging the entire cable, except for the two ends, in water. Set your Ohmmeter to RX100K, and adjust it to zero. Put one of the meter leads on a cable wire and the other to a ground. Test each wire in the cable with this same procedure.

If at anytime the meter's needle goes to zero, remove the splice connection from the water and watch the needle. If the needle falls back to give no reading, the leak is in the splice.

Once the splice is ruled out, you have to test sections of the cable in a similar manner. In other words, once you have activity on the meter, you should slowly remove sections of the cable until the meter settles back into a no-reading position. When this happens, you have found the section that is defective. At this point, the leak can be covered with waterproof electrical tape and reinstalled, or you can replace the cable.

Will Not Run

A pump that will not run can require extensive troubleshooting. Start with the obvious, and make sure the fuse is not blown and the circuit breaker is not tripped. Also check to see that the fuse is of the proper size.

Incorrect voltage can cause a pump to fail. You can check the voltage as described in the previous electrical troubleshooting section.

Loose connections, damaged cable insulation, and bad splices, as discussed earlier, can prevent a pump from running.

The control box can have a lot of influence on whether or not a pump will run. If the wrong control box has been installed or if the box is located in an area where temperatures rise to over 122 degrees Fahrenheit, the pump might not run.

When a pump will not run, you should check the control box out carefully. We will be working with a quick-disconnect type box. Start by checking the capacitor with an Ohmmeter. First, discharge the capacitor before testing. You can do this by putting the metal end of a screwdriver between the capacitor's clips. Set the meter to RX1000, and connect the leads to the black and orange wires out of the capacitor case. You should see the needle start moving toward zero and then swing back to infinity. Should you have to recheck the capacitor, reverse the Ohmmeter leads.

The next check involves the relay coil. If the box has a potential relay (three terminals), set your meter on RX1000 and connect the leads to red and yellow wires. The reading should be between 700 and 1800 Ohms for 115-volt boxes. A 230-volt box should read between 4500 and 7000 Ohms.

If the box has a current relay coil (four terminals), set the meter on RX1 and connect the leads to black wires at terminals one and three. The reading should be less than one Ohm.

To check the contact points, you will set your meter on RX1 and connect to the orange and red wires in a three-terminal box. The reading should be zero. For a four-terminal box, you will set the meter at RX1000 and connect to the orange and red wires. The reading should be near infinity.

Now you are ready to check the overload protector with your Ohmmeter. Set the meter at RX1, and connect the leads to the black wire and the blue wire. The reading should be at a maximum of 0.5.

If you are checking the overload protector for a control box designed for 1½ horsepower or more, you will set your meter at RX1 and connect the leads to terminal number one and to terminal number three on each overload protector. The maximum reading should not exceed 0.5 Ohms.

A defective pressure switch or an obstruction in the tubing and fittings for the pressure switch could cause the pump not to run.

As a final option, the pump might have to be pulled and checked to see if it is bound. There should be a high amperage reading if this is the case.

Doesn't Produce Water

When a submersible pump runs but doesn't produce water, there are several things that could be wrong. The first thing to determine is if the pump is submerged in water. If you find that the pump is submerged, you must begin your regular troubleshooting.

Loose connections or wires connected incorrectly in the control box could be at fault. The problem could be related to the voltage. A leak in the piping system could easily cause the pump to run without producing adequate water.

A check valve could be stuck in the closed position. If the pump was just installed, the check valve might be installed backwards. Other options include a worn pump or motor, a clogged suction screen or impeller, and a broken pump shaft or coupling. You will have to pull the pump if any of these options are suspected.

TANK PRESSURE

If you don't have enough tank pressure, check the setting on the pressure switch. If that's okay, check the voltage. Next, check for leaks in the piping system, and as a last resort, check the pump for excessive wear.

FREQUENT CYCLING

Frequent cycling is often caused by a waterlogged tank, as was described in the section on jet pumps. Of course, an improper setting on the pressure switch can cause a pump to cut on too often, and leaks in the piping can be responsible for the trouble. You might find the problem is being caused by a check valve that has stuck in an open position.

Occasionally the pressure tank will be sized improperly and cause problems. The tank should allow a minimum of one minute of running time for each cycle.

WATER QUALITY

Water quality in private water supplies can vary greatly from one building lot to another. It can even fluctuate within one water source. A well that tests fine one month might test differently six months later. Frequent testing is the only way to assure a good quality of water. Many times, some type of treatment is either desirable or needed before water can be used for domestic purposes.

Many people have heard of hard water and soft water. While these names might be familiar, a lot of people don't know the differences between the two types of water. Some water supplies have hazardous concentrations of contaminants, but most wells are affected more often by mineral contents that, while not necessarily harmful to a person's health, can create other problems.

Is it a builder's responsibility to provide customers with water that is free of mineral content? It shouldn't be. Water conditioning equipment can get very expensive. Spending $1500 or more for such equipment would not raise any eyebrows in the plumbing community. If you want to make sure that you don't become entangled in a lot of litigation over water quality, have your lawyer create a disclaimer clause for your contracts.

There are four prime substances that can affect the quality of water. Physical characteristics are the first of the four substances. By physical characteristics, we are talking about such factors as color, turbidity, taste, odor, and so forth. Chemical differences create a second category. This type of situation can involve such topics as hard and soft water. Biological contents can modify both the physical and chemical characteristics of water. This third factor in determining water quality can often render water unsafe for drinking. The fourth consideration is radiological factors, such as radon. To understand these four groups better, let's look at them individually.

Physical Characteristics

Examining the physical characteristics of water is one way of assessing water quality. Taste and odor rank high as concerns in the category of physical characteristics. Water that doesn't taste good or that smells bad is undesirable. It might be perfectly safe to drink, but its physical qualities make it unpleasant to live with.

What causes problems with taste and odors? Foreign matter is at fault. It might come in the form of organic compounds, inorganic salts, or dissolved gases. A sulfur content is one of the best known causes of odor in water.

Water that doesn't look good can be difficult to drink enjoyably. Color is a physical characteristic that can taint someone's opinion of water quality. Having colored water rarely indicates a health concern, but it is a sure way of drawing some customer complaints. Dissolved organic matter from decaying vegetation and certain inorganic matter typically give water color.

Turbidity. Do you know what turbid water is? It's simply water that is cloudy in appearance. Turbidity is caused by suspended materials in the water. Such materials might be clay, silt, very small organic material, plankton, and inorganic materials. Turbidity doesn't usually pose a health risk, but it makes water distasteful to look at in a drinking glass.

Temperature. Technically, temperature falls into the category of physical characteristics. This, however, is not a problem that many people complain about. Deep wells produce water at consistent temperatures. Shallow wells are more likely to have water temperatures that fluctuate. In either case, temperature is rarely a concern with wells.

Foam. Foam is something that you will not normally encounter in well water. However, foamability is a physical characteristic of water, and it can be an indicator that serious water problems exist. Water that foams is usually being affected by some concentration of detergents. While the foam itself might not be dangerous, the fact that detergents are reaching a water source should raise some alarm. If detergents can invade the water, more dangerous elements might also be present.

Chemical Characteristics

Chemical characteristics are often monitored in well water. A multitude of chemical solutions might be present in a well. As long as quantities are within acceptable, safe guidelines, the presence of chemicals does not automatically require action. However, concentrations of some chemicals can prove harmful. The following is a list of some chemicals that might be present in the next well you have installed:

- Arsenic

- Barium

- Cadmium

- Chromium

- Cyanides

- Fluoride

- Lead

- Selenium

- Silver

Chlorides.   Chlorides, in solution, are often present in well water. If an excessive quantity of chlorides is present, it might indicate pollution of the water source.

Copper.   Copper can be found in some wells as a natural element. Aside from giving water a poor taste, small amounts of copper are not usually considered harmful.

Fluorides.   Why pay a dentist to give your children fluoride treatments when fluorides might be present in your drinking water. Natural fluorides can be found in some well water. Too much fluoride in drinking water is not good for teeth, so quantities should be measured and assessed by experts.

Iron.   Iron is a common substance found in well water. If water has a high iron content, it will be difficult to avoid brown stains on freshly washed laundry. Plumbing fixtures can be stained by water containing too much iron. The taste of water containing iron can be objectionable.

Lead.   Lead is one thing you don't want to have show up on a well test. It's possible for dangerous levels of lead to exist in a water source, but most water containing lead derives the detrimental element from plumbing pipes. Modern plumbing codes have provisions to guard against lead being introduced to potable water from piping, but older homes don't share this safeguard.

Manganese.   Manganese, like iron, is very common in well water. Staining of laundry and plumbing fixtures is one reason to limit the amount of manganese in domestic water. It is not unusual for manganese to affect the taste of water adversely. Additionally, excessive consumption of manganese can cause health problems.

Nitrates. Nitrates show up most often in shallow wells. Nitrate can cause what is known as blue-baby disease in infants who ingest water containing it. Shallow wells that are located near livestock are susceptible to nitrate invasion.

Pesticides. As we all know, pesticides and well water don't mix—at least they shouldn't. Shallow wells located in areas where pesticides are used should be checked often to confirm the suitability of the well's water for domestic use. Many reports exist that indicate wells have been contaminated during ground treatments for termite control around houses.

Sodium. Sodium can show up in well water. For average, healthy people, this is not a problem. However, individuals who are forced to maintain low-sodium diets can be affected by the sodium content in well water.

Sulfates. Sulfates in well water can act as a natural laxative. You can imagine why this condition would not be desirable in most homes.

Zinc. Zinc doesn't normally draw attention to itself as a health risk, and it is not a common substance in well water. However, it can sometimes be present. Taste is normally the only objection to a content of zinc in well water.

Hard Water. Hard water is a common problem among well users. This type of water doesn't work well with soap and detergents. If you heat a pot of water on a stove and find a coating of white dustlike substance left in the pan, it is a strong indication of hard water being present. This same basic coating can attack plumbing pipes and storage tanks, creating a number of plumbing problems. Hard water can even cause flush holes in the rims of toilets to clog up and make the toilets flush slowly or poorly.

Acidic Water. Acidic water is not uncommon in wells. When water has a high acid content, it can eat holes through the copper tubing used in plumbing systems. Plumbing fixtures can be damaged from acidic water. People with sensitive stomachs can suffer from a high acid content. The acidity of water is measured on a pH scale. This scale runs from 0 to 14. A reading of 7 indicates neutral water. Any reading below 7 is moving into an acidic range. Numbers above 7 are inching into an alkaline status.

Biological Factors

Biological factors can be a big concern when it comes to well water. To call water potable or suitable for domestic use, it must be free of disease-producing organisms. What are some of the organisms? Bacteria, especially of the coliform group, is one of them. Protozoa, virus, and helminths (worms) are others.

Biological problems can be avoided in many ways. One way is to use a water source that doesn't support much plant or animal life, such as a well.

Springs, ponds, lakes, and streams are more likely to produce biological problems than covered wells are.

It is also necessary to protect a potable water source from contamination. The casing around a well does this. Light should not be able to shine on a water source, and a well cap or cover meets this requirement. Temperature can also play a part in bacterial growth, but the temperature of most wells is not something to be concerned about. If biological activity is a problem, various treatments to the water can solve the problem.

Radiological Factors

Radiological factors are not a big threat to most well users, but some risk does exist. Special testing can be done to determine if radioactive materials are a significant health risk in any given well. This test should, in my opinion, be conducted by experienced professionals. The same goes for biological testing.

SOLVING WATER-QUALITY PROBLEMS

There are enough ways of solving water-quality problems that a small book could be written on the subject. Rather than give you a full tour of all aspects of water treatment, I will concentrate on the methods most often used in average homes. There is some form of treatment available for nearly any problem you encounter.

Bacteria

Bacteria in well water is serious. The most common method of dealing with this problem doesn't require any fancy treatment equipment. A quantity of chlorine bleach is usually all that is needed. Bleach is added to the contaminated well water and allowed to settle for awhile. After a prescribed time, the well is drained or run until no trace of the bleach is evident. This normally clears up biological activity. Sophisticated treatment systems do exist for nasty water, but the odds of needing one are remote.

Acid Neutralizers

Acid neutralizers are available at a reasonable cost to control high acid contents in domestic water. These units are fairly small, easy to install, not difficult to maintain, and don't cost a small fortune to purchase.

Iron

Iron and manganese can be controlled with iron-removal filters. Like acid neutralizers, these units are not extremely expensive, and they can be installed in a relatively small area. If both a water softener and iron-removal system are needed, the iron-removal system should treat water before it reaches a water softener. Otherwise, the iron or manganese might foul the mineral bed in the water softener.

Water Softeners

Water softeners can treat hard water and bring it back to a satisfactory condition. The use of such a treatment system can prolong the life of plumbing equipment, while providing users with more desirable water quality.

Activated Carbon Filters

Activated carbon filters provide a solution for water that has a foul taste or odor. Many of these filters are simple, in-line units that are inexpensive and easy to install. When conditions are severe, a more extensive type of activated carbon filter might be required. This type of filter can remove the ill effects of sulfur water (hydrogen sulfide).

Turbidity

Turbidity can be controlled with simple, in-line filters. If the water being treated contains high amounts of particles, the cartridges in these filters will have to be changed frequently. Left unchanged, they can collect so many particles that water pressure is reduced greatly.

CHAPTER 5
PRIVATE SEWAGE DISPOSAL SYSTEMS

There are site limitations for private sewage disposal systems. Not all pieces of land are suitable for such systems. Land developers and builders must be able to spot land that is likely to give them trouble when it comes to a septic system. Experience does much to help in this area. After several years of buying and developing land, a person begins to know good soil when its seen.

There are little signs that can give you hints about the quality of land. For example, bulges and occasional glimpses of rock on the land's surface could mean that bedrock is close to the surface. This will certainly interfere with a private sewerage disposal system.

Rock is not the only risk when it comes to septic systems. Some ground just won't perk. If this is the case, the land might be deemed unbuildable. Buying land for development and then discovering that there are no acceptable septic sites can really ruin your day. Experienced land buyers have clauses in their purchase agreements to protect them from this type of risk. A typical land agreement will contain a contingency clause that gives the buyer a chance to have the soil tested before an absolute commitment is made to buy the land. If the tests prove favorable, the deal goes through. When soils studies (Figure 5.1) turn up problems, the contract might be voided or some compensation might be made in the sales price.

	Slow absorption	Medium absorption	Rapid absorption
Seconds required for water to drop 1 inch	5–30	3–5	0–3

FIGURE 5.1 Soil absorption ratings.

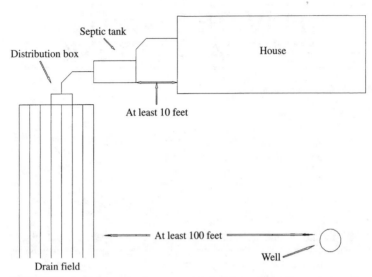

FIGURE 5.2   Septic layout showing recommended minimum distance from a water well.

The size of a building lot can affect its ability to be approved for a septic system. Many houses that require a septic system for sewage disposal also require a water well for drinking water. For obvious reasons, septic systems must not be installed too closely to a water well (Figure 5.2). The minimum distance between the two is normally 100 feet, or more. I have seen some exceptions to this rule, but not many.

I've run into a number of building lots where municipal water hook-ups were available, but where a private waste disposal system was needed. With my contingency clauses and inspections, I've never been put in a bind from this type of thing. However, a buyer or builder who doesn't research what options are available could get in big trouble very quickly.

Some people assume that, since building lots on either side of a particular lot are able to use septic systems, the lot in question should be suitable for such a system. This is not always the case. I've seen land where, out of five acres, there might be only one or two sites suitable for a standard septic system. This situation is not uncommon, so watch out for it. Make sure that you have an approved septic location established before you make any firm commitments to buy or build.

GRADE

The grade of a building lot can have a lot to do with the type of septic system that must be installed. If the grade will not allow for a gravity system, the price that you pay for a septic system will go up considerably. Pump stations can be used, but they are not inexpensive. If you're building houses on spec, you might have trouble selling one that relies on a pump station. People don't like the idea of having to replace pumps at some time in the future. Many people are afraid the pumps will fail, leaving them without sanitary conditions until the pump can be replaced.

The naked eye is a natural wonder, but it cannot always detect the slope of a piece of land accurately. Looks can be deceiving. Unless the land you are looking at leaves no room for doubt in its elevation, check the land with a transit. You can't afford to bid jobs with regular septic systems and then wind up having to install pump systems.

A SAFE WAY

There is a safe way to work with land where a septic system is needed. You can go by reports that are provided to you by experts. Even this is not foolproof, but it's as good as it gets. If you have a soils engineer design a septic system for your job, you can be pretty sure that the system will work out close to the way it has been drawn. Because every jurisdiction I know of requires a septic design before a building permit will be issued, it makes sense to go ahead and get the design early.

If you are buying land, make sure that your purchase agreement provides a contingency that will allow you to have the land approved for a septic system before you are committed to going through with the sale. It is wise to specify in your contingency what type or types of septic designs you will accept. Some types of systems cost much more than others. If your contract merely states that the land must be suitable for a septic system, you will have to complete the sale regardless of what the cost of a system is, so long as one can be installed.

When you are bidding jobs, make clear in your quote what your limits are in regards to site conditions. Specify the type of septic system your quote is based on, and provide language that will protect you if some other type of system becomes required. Don't take any chances when dealing with septic systems, because the money you lose can be significant.

A STANDARD SYSTEM

A standard septic system (Figure 5.3) will be built with a gravity flow. Its drain field, or *leach field* as it is often called, will be made up of perforated pipe and crushed stone. This is the least expensive type of septic system to install. Unless the soil will not perk well, you will normally be able to use a standard system.

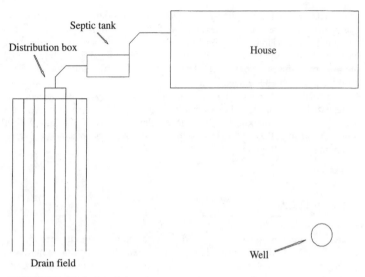

FIGURE 5.3 Typical septic layout.

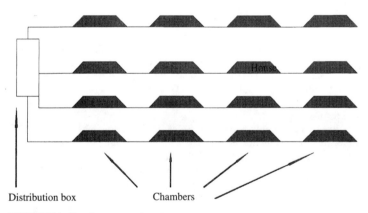

FIGURE 5.4 Chamber type septic system.

CHAMBER SYSTEMS

Chamber systems (Figure 5.4) are much more expensive than pipe-and-gravel systems. When soil doesn't perk well, chamber systems are used. If the soil will perk, but not perk well, a chamber system might be your only choice.

How do these systems work? Basically, the chambers hold effluent from a septic tank until the ground can absorb it. Unlike a perforated pipe, which would release effluent quickly, the chamber controls the flow of effluent at a rate acceptable to the soil conditions. Where a pipe-and-gravel system might flood an area with effluent, a chamber system can distribute the liquid more slowly and under controlled circumstances. The cost of a chamber system can easily be twice that of a pipe-and-gravel system.

PUMP STATIONS

Pump stations (Figure 5.5) are another big expense in some septic systems. The cost of the pump station, the pump and its control, and the additional labor required to install such a system can add thousands of dollars to the cost of a standard septic system. If you were unlucky enough to get stuck with a pumped chamber system when you had planned on a gravity gravel system, you could lose much of your building profit all at once.

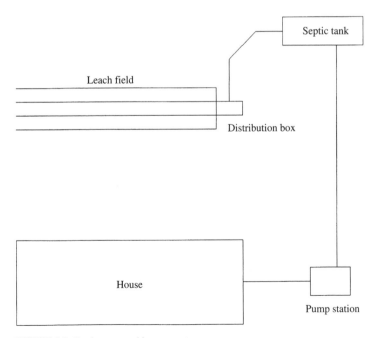

FIGURE 5.5 Septic system with pump system.

TREES

Trees are another factor you must consider when doing a site inspection for a septic system. Tree roots and drain fields don't mix very well. If there are trees in the area of the septic system, they must be removed. Even trees that are not directly in the septic site should be removed from the edges of the area.

How far should open space exist between a septic system and trees? It depends on the types of trees that are growing in the area. Some trees have roots that reach out much farther than others. You can get a good idea of how far a tree's roots extend by looking that the branches on the tree. The spread of the branches will often be similar to the spread of the roots. If there is any doubt, ask the professional who draws your septic design to advise you on which trees can be left standing.

BURYING A SEPTIC TANK

Burying a septic tank (Figure 5.6) requires a fairly deep hole. Even if you are using a low-profile tank, the depth requirement will be several feet. If you are working in an area where bedrock is present you must be cautious. You could run into a situation where rock will prevent you from burying a septic tank.

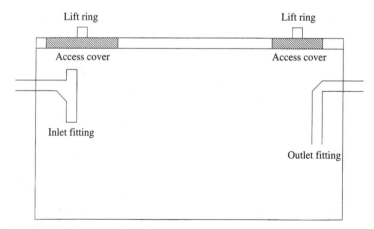

FIGURE 5.6   Common septic tank.

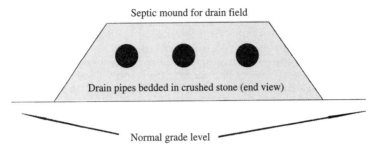

Septic mound for drain field

Drain pipes bedded in crushed stone (end view)

Normal grade level

FIGURE 5.7 Mound-type septic system.

UNDERGROUND WATER

Underground water can present problems for the installation of a septic system (Figure 5.7). This type of problem should be detected when test pits are dug for perk tests. However, it is possible that the path of the water would evade detection until full-scale excavation was started. For this reason, you should have some type of language in your contracts with customers to indemnify you against underground obstacles, such as water.

DRIVEWAYS AND PARKING AREAS

When you assess a lot for a septic system, consider the placement of driveways and parking areas. Even though a septic system will be below ground, it is not wise to drive vehicles over the system. The weight and movement can damage the drain field to a point where replacement is required. You certainly don't need this type of warranty work, so make sure that all vehicular traffic will avoid the septic system.

EROSION

Erosion can be a problem with some building lots and land. If you install a septic field on the side of a hill, you must make sure that the soil covering the field will remain in place. This can be done by planting grass or some other ground cover. However, when you check out a piece of land, you need to take the erosion factor into consideration. The cost of preventing a wash-out over the septic system could add a significant amount of expense to your job.

SET-BACKS

Set-backs are something else that you should check on before committing to a septic design. Many localities require all improvements made on a piece of land to be kept a minimum distance from the property lines. A typical set-back for a side property line is 15 feet, but this is not always the case. Where I live, there are no set-backs. However, I've seen set-back requirements that were more than 15 feet. This can become a very big factor in the installation of a septic system.

COMMON PRACTICES

There are many common practices associated with septic systems. Generally, all specifications for a system will be determined by some expert and given to you or your subcontractor. For example, the size of a septic tank (Figure 5.8) is based on the number of bedrooms in a home. You might think the size would be based on bathrooms, but it is not.

When you contract with your plumber to hook up a septic system, make sure that short-turn fittings (Figure 5.9) will not be used. Long-turn fittings should be used. Plumbing codes require this in most areas, but it never hurts to put specifics in writing. An above-ground clean-out should also be installed just outside of the building foundation (Figure 5.10).

Septic tank systems that are installed properly involve a sewer, a septic tank, a distribution box, and a drain field (Figure 5.11). A tank that is in good

Single family dwellings, number of bedrooms	Multiple dwelling units or apartments, one bedroom each	Other uses; Maximum fixture units served	Minimum septic tank capacity in gallons
1–3		20	1000
4	2	25	1200
5 or 6	3	33	1500
7 or 8	4	45	2000
	5	55	2250
	6	60	2500
	7	70	2750
	8	80	3000
	9	90	3250
	10	100	3500

FIGURE 5.8   Common septic tank capacities.

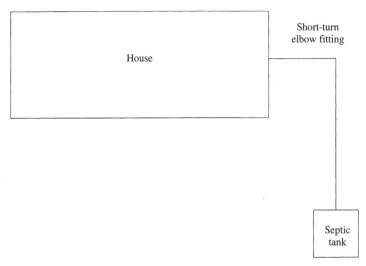

FIGURE 5.9 Septic system using short-turn elbow fitting.

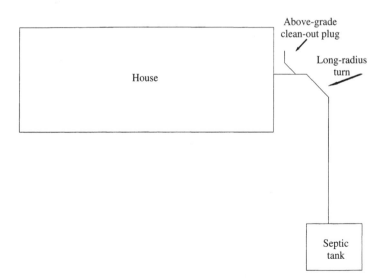

FIGURE 5.10 Septic system using long-turn elbow fitting.

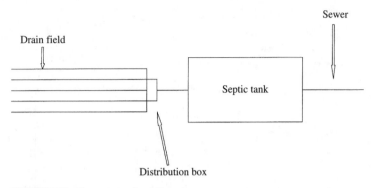

FIGURE 5.11 Components of a septic system.

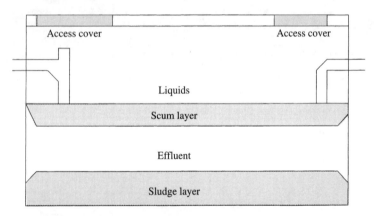

FIGURE 5.12 Layers of a typical septic system.

operating condition will contain a sludge layer at the bottom, effluent above it, and a scum layer near the top (Figure 5.12). Septic professionals can measure these components to determine when a tank needs to be pumped. If you have questions pertaining to septic systems make sure you check with a professional before committing to a decision.

CHAPTER 6
FOOTINGS AND FOUNDATION WALLS

Footings and foundation walls are elements of construction that builders and remodelers have to work with on a regular basis. You might be installing a pier foundation for a deck, installing a full foundation for a house, or sealing a foundation to prevent water leakage. Bedrock, or *ledge* as it is often called, can provide a footing for a foundation. However, footings are generally made of concrete that is poured in a hole or trench that is at least 6 inches below the local frost line. Let's start our discussion of foundations with ledge.

LEDGE

In parts of the country, such as Maine, there is a lot of rock. When it is bedrock, it is sometimes called ledge. This ledge can be deep in the ground, or it can protrude up above ground level. It is frequently only a few feet below the grade level. This situation can ruin plans for a full basement. The only way to deal with ledge is to blast it out, and this gets extremely expensive. However, most code authorities recognize bedrock as a natural footing that can be built on. Check with your local code officers before making this assumption, but you will probably find that bedrock can be used as a footing.

TYPES OF FOOTINGS

There are two basic types of footings for foundations. *Pier footings* are created by digging holes in the ground and filling them with concrete. Support posts later rest on the concrete in the holes and provide a stable foundation for a building (Figure 6.1). It is common for preformed tubes to be used when pier foundations are made. The tubes resist frost heaves and, therefore, result

---

- Place piers 8 feet on center when positioned perpendicular to floor joists.
- Place piers 12 feet on center when positioned parallel to floor joists.

FIGURE 6.1 Standard pier foundation spacing (one-story residence in average soil).

Material	Support capability
Hard rock	80 tons
Loose rock	20 tons
Hardpan	10 tons
Gravel	6 tons
Coarse dry sand	3 tons
Hard clay	4 tons
Fine dry sand	3 tons
Mixed sand and clay	2 tons
Wet sand	2 tons
Firm clay	2 tons
Soft clay	1 ton

FIGURE 6.2 Bearing capacities of foundation soils.

Foundation footings should be installed at least 1 foot below the local frost-line level. This can result in a wide variation of depth, depending upon the geographical area in which you are working. Some footings, such as in Maine, must be 4 feet deep, while footings in warmer climates might be only 18 inches deep. Consult your local code book or code officer for an exact depth to install footings in your area.

FIGURE 6.3 Footing depths.

Load divided by soil-bearing capacity will reveal the required number of square feet, or area, required of a foundation footing. For example, if you had a load of 60 tons per square foot that would be built in soil with a capacity rating of 20, you would need 3 square feet of footing. This means that your footing would be 3 feet wide.

FIGURE 6.4 Finding the area of a footing.

in less seasonal movement than what would occur if the concrete was poured directly into a hole in the earth (Figure 6.2).

Continuous or running footings are created by digging a trench and filling it with concrete (Figures 6.3 and 6.4). The trench provides a full boundary for the structure being built. When a full foundation wall is wanted, a continuous footing is needed. All footings should be installed at least 6 inches below the local frost level.

Continuous footings are normally used when full-size structures, such as homes, are being built. Pier foundations are frequently used for decks and porches. However, there are many homes that are supported on pier foundations. Because pier foundations are less expensive than full foundations, they are often used on camps, cottages, decks, porches, and other buildings where a full foundation is not required.

TYPES OF FOUNDATIONS

There are many types of foundations. Each type has its place in construction, and some are better than others, depending upon their uses. Remodelers are often hired to add living space onto homes. This involves the installation of new foundations. Choosing the best, and most cost-effective, type of foundation for your jobs will make your bids more competitive and your customers happier.

Slab Foundations

Slab foundations (Figures 6.5 through 6.7) are about the least expensive type to use for living space. This doesn't necessarily mean they are the best, but they usually cost the least. Because a slab foundation gives you footings and a subfloor all in one package, the cost of constructing a room addition on a slab is quite affordable, when compared to other foundation options.

Slab thickness (inches)	Slab area (square feet)				
	10	**50**	**100**	**300**	**500**
2	0.1	0.3	0.6	1.9	3.1
3	0.1	0.5	0.9	2.8	4.7
4	0.1	0.6	1.2	3.7	6.2
5	0.2	0.7	1.5	4.7	7.2
6	0.2	0.9	1.9	5.6	9.3

FIGURE 6.5 Estimating cubic yards of concrete for slabs, walks, and drives.

Use	Thickness
Basement floors in residences	4 inches
Garage floors for residential use	4 to 5 inches
Porch floors	4 to 5 inches
Base for tile flooring	2½ inches
Driveways	6 to 8 inches
Sidewalks	4 to 6 inches

FIGURE 6.6 Thickness of a concrete slab.

Multiply the length by the width by the thickness. For example, a project that is 50 feet long, 10 feet wide, and 8 inches deep will require 333.33 cubic feet of concrete. Because concrete is sold in yards, convert your findings by dividing the total (333.33) by 27. You will arrive at an answer of 12.35 cubic yards.

FIGURE 6.7 Estimating concrete volume.

There are some disadvantages to slab foundations. For one thing, underground installations, such as plumbing, electrical wiring, and heating and air-conditioning work are not accessible once the floor is poured. Some people don't mind this, but others can't stand the idea of not being able to get to their mechanical installations.

Slab floors tend to be cold. Customers might not think of this when planning a job, but they will think of it when they walk around on the finished floor without shoes on. Carpeting helps to combat the cold, but there is a noticeable difference between a concrete floor and a wood floor. There is also the issue of concrete floors being harder than wood floors.

Moisture sometimes seeps into slab floors. This isn't a routine problem, but it can occur. If moisture invades a slab, it can cause floor covering to come unglued. It can also make carpeting damp, musty, and in the long term, it can destroy the floor covering.

Storage under a slab floor is nonexistent. There is no room for mechanical equipment, such as a heat pump or water heater. This might not be much of a factor in remodeling work, but it is something to consider when building a new home.

Pier Foundations

Pier foundations can be very cost-effective for certain types of projects. They are an obvious choice for decks, but they can be used for several types of additions. Screen porches do very well when built on pier foundations. Sun rooms can also be built on piers. Other types of rooms can be placed on piers, but they generally look more out of place and present some difficulties with mechanical installations.

If the room being built on piers is not equipped with mechanical installations, there is little need to block off the underside of the addition. I'm not saying that the floor joists shouldn't be insulated, but there will be no need to protect pipes from freezing or air ducts from exposure. Installing lattice around the perimeter of the piers will enclose the foundation to enhance its eye appeal.

When plumbing, heating, or air ducts are installed under the structure that is built on piers, additional precautions might need to be taken. Because exposure to outside elements often affects the performance of these systems, you might have to build an enclosed chaseway under the addition to protect the systems. This is still much less expensive than constructing a full foundation.

Crawl Spaces

Houses built on crawl spaces are common (Figure 6.8). There should be at least 18 inches of height between the ground and any floor joists; however, other than for that, a crawl space can have as much height as is needed. Crawl-space foundations cost much less than full basements to build. Additions built over crawl spaces make it relatively easy to install mechanical systems, and the enclosed foundation protects the systems from outside elements.

Having a crawl space foundation allows access to mechanical systems and framing systems. This gives a lot of customers peace of mind. They know if anything ever goes wrong under their floor, it can be gotten to. If new mechanical equipment, such as a heating system, is required to handle a new addition, it can often be tucked into the crawl space.

From an appearance point of view, crawl-space foundations normally beat out slabs and piers. The exterior can be veneered with brick (Figures 6.9

	Multiply free vent area by	
Vent cover material	**With soil cover**	**No soil cover**
¼" mesh hardware cloth	1.0	10
⅛" mesh screen	1.25	12.5
16-mesh insect screen	2.0	20
Louvers + ¼" hardware cloth	2.0	20
Louvers + ⅛" mesh screen	2.25	22.5
Louvers + 16-mesh screen	3.0	30

FIGURE 6.8 Crawl space gross vent area requirements.

Brick type	Height (in.)	Width (in.)	Length (in.)
Standard building brick	2½	3⅞	8½
Oversize building brick	3¼	3¼	10
Norman face brick	2³⁄₁₆	3½	11½
Fire brick	2½	4½	9
Fire brick splits	1¼	4½	9

FIGURE 6.9 Brick sizes.

- Cavity wall
- Brick veneer
- Single wythe
- Composite brick

FIGURE 6.10 Types of brick walls.

Material	Height	Units	Cement	Sand
Standard brick	4-inch wall	616	3 sacks	9 cubic feet
Standard bricks	8-inch wall	1232	7 sacks	21 cubic feet
Standard block	8-inch wall	112	1 sack	3 cubic feet

FIGURE 6.11 Figuring square footage needs for a 100-square-foot masonry wall.

- Stretcher
- Rowlock stretcher
- Header
- Rowlock header
- Soldier

FIGURE 6.12 Types of bricks.

through Figure 6.18), or it can be swirled with a stucco pattern and painted. Unlike a slab, where exterior siding must be installed low to the ground, additions built on crawl spaces can have their siding start well above ground level, avoiding some moisture problems. Most room additions work very well on crawl-space foundations.

- Flemish
- Flemish Cross
- Garden
- Common
- Common with Flemish headers
- Running
- English
- Dutch

FIGURE 6.13 Types of brick wall construction.

To create 8 square feet of block wall, you will need nine 8-×-8-×-16 blocks. The formula for this equation is simple. Take the square footage of your proposed wall and multiply it by 1⅛ (1.125) to determine your material needs.

FIGURE 6.14 Estimating block needs.

Not weather-resistant
- Flush
- Raked
- Struck

Weather-resistant
- Concave
- V-shaped
- Weathered

FIGURE 6.15 Facts about mortar joints.

Mortar	Rated for
Type M	Vigorous exposure, load-bearing capability, and below-grade use
Type S	Severe exposure, load-bearing capability, and below-grade use
Type N	Light loads, mild exposure, and above-grade use
Type O	Light loads and interior use
Type K	Nonbearing use or for where compressive strength does not exceed 75 PSI

FIGURE 6.16 Mortar ratings.

Mortar	Compressive strength (PSI)
Type M	2500
Type S	1800
Type N	750
Type O	350
Type K	75

FIGURE 6.17 Mortar statistics.

Grade	Ratio	Material needed for cu. yd.
Strong—watertight, exposed to weather and moderate wear	1:2 ¼:3	6 bags cement 14 cu. ft. sand (.52 yd³) 18 cu. ft. stone (.67 yd³)
Moderate—Strength, not exposed	1:2 ¾:4	5 bags cement 14 cu. ft. sand (.52 yd³) 20 cu. ft. stone (.74 yd³)
Economy—Massive areas, low strength	1:3:5	4½ bags cement 13 cu. ft. sand (.48 yd³) 22 cu. ft. stone (.82 yd³)

FIGURE 6.18 Concrete formulas.

Basements

Basements are nice from several standpoints. They make installing mechanical equipment easy. Access to building components is excellent in a basement, and an unfinished basement provides significant room for storage. Most people never seem to have enough storage, so this is a major selling point for a basement.

The drawbacks to basements include possible moisture problems and considerable expense during construction. Some basements fill with water on a regular basis. Subsoil drains, sumps, draintile, and sump pumps can be installed to correct most water problems, but the effort and expense has to be accounted for. Excavation for a basement gets expensive, and so does the additional height of the foundation walls. If money is no object, then a basement is rarely a bad choice.

CHAPTER 7
FLOORING SYSTEMS

Floor systems in homes are not very complicated. They are made up of plates, bands, girders, joists, and sheathing. There might be some bridging, and there will be nails. Compared to some parts of a job, floor systems are actually simple. This doesn't mean, however, that they don't provide a level of risk.

If you are doing a standard interior remodel, the floor system might not have to be disturbed. In fact, it probably won't. It is not uncommon to replace some subflooring or to add some underlayment, but major alterations in a floor system are not normally part of an average remodeling job. Attic conversions are an exception to this rule, and some basement conversions require the construction of a flooring system. However, on the whole, structural flooring is not a key element in remodeling. Builders of new construction do, however, deal with flooring systems regularly.

ATTIC CONVERSIONS

Attic conversions are one type of remodeling where major work is usually required with a floor system. When space is designed for attic use, the structural members are not rated for live loads (Figures 7.1 through 7.5). For example, 2-×-6 lumber might be used in the framing of an attic floor, where 2-×-8, or larger, dimensional lumber would be required for a live load. Not only is the lumber for an attic usually smaller, it is often spaced on 24-inch centers, where spacing for habitable space is normally done on 16-inch centers. The combination of undersized lumber and extended spacing makes it necessary for substantial framing to be done before the attic can be used as living space.

Whenever you are getting involved with major structural issues, such as upgrading an attic floor for use as living space, experts should be consulted for design information. If you take it upon yourself to select the size and spac-

- The first floor of a residential dwelling should be rated at 40 pounds per square foot (PSF).
- Other floors should be rated at 30 PSF.
- Stair treads should be rated at 75 PSF.
- Roofs used for sun decks should be rated at 30 PSF.
- Garages for passenger cars should be rated at 50 PSF.
- Attics accessible by stairs or ladder should be rated at 30 PSF, when the ceiling height is more than 4½ feet.
- Attics accessible by scuttle hole with a ceiling height of less than 4½ feet can be rated at 20 PSF.

FIGURE 7.1 Minimum uniformly distributed live loads.

Joist size (inches)	Spacing (inches)	Pine/fir (feet/inches)
2 × 6	12	10-6
2 × 6	16	9-8
2 × 6	24	8-4
2 × 8	12	14-4
2 × 8	16	13-0
2 × 8	24	10-4
2 × 10	12	17-4
2 × 10	16	16-2
2 × 10	24	14-6
2 × 12	12	20-0
2 × 12	16	18-8
2 × 12	24	16-10

FIGURE 7.2 Typical maximum floor joist spans with 40-pound live load.

Area/activity	Live load, PSF
First floor	40
Second floor and habitable attics	30
Balconies, fire escapes, and stairs	100
Garages	50

FIGURE 7.3 Residential live loads.

To find a dead-load rate, you have to calculate all building materials and their effective weight ratios. For example, if you have a building that has 2-×-8 floor joists, set 16 inches on center, that are covered with ½"-plywood and carpet, you have a dead load of 5 PSF. This solution is arrived at by taking the rating of the joists (3 PSF) plus the weight of the plywood (1.5 PSF) and the carpeting (0.5 PSF).

FIGURE 7.4 Dead load weight determination.

Material	Load per PSF
Standard 2×4 (16" on center)	2
Standard 2×6 (16" on center)	2
Standard 2×8 (16" on center)	3
Standard 2×10 (16" on center)	3
Softwood, per inch of thickness	3
Hardwood, per inch of thickness	4
Plywood, per inch of thickness	3
Concrete, per inch of thickness	12
Stone, per inch of thickness	13
Carpet, per inch of thickness	0.5

FIGURE 7.5 Dead load weights of various flooring materials.

ing of new floor joists, you are assuming a liability that, in my opinion, is better left to engineers and architects. My best advice to you is to seek help from qualified design professionals and then follow their instructions.

While I am not an architect or engineer, I can tell you how most of the attic conversions I've worked with have been done. Don't take these words as your excuse to bypass the help of local design professionals; accept it as background information.

First of all, if you are going to convert an attic into living space, prepare your customers well for the events that will unfold during the conversion (Figure 7.6). For example, tell them how the existing ceiling that is attached to the structural members in the attic will be damaged. It is all but impossible to complete an attic conversion without causing some degree of damage to the ceiling below. Once the customer has been told that ceiling light fixtures should have their covers removed, that their ceilings might develop cracks and nail-pops, and that dust and dirt is likely to filter down through the ceiling, you are ready to begin the physical work.

Joist size (inches)	Spacing (inches)	Pine (feet/inches)
2 × 6	12	11-10
2 × 6	16	10-10
2 × 6	24	9-6
2 × 8	12	17-2
2 × 8	16	16-0
2 × 8	24	14-4
2 × 10	12	21-8
2 × 10	16	20-2
2 × 10	24	18-4
2 × 12	12	24-0
2 × 12	16	23-10
2 × 12	24	21-10

FIGURE 7.6 Typical maximum ceiling joist spans (no attic storage capacity).

Assuming that you are working with a stick-built roof, you will have rafters, collar ties, and bottom cords to work around. For the time being, it is the bottom cords that we are interested in. There is no reason why these pieces of lumber should have to be removed. You can simply install new floor joists around them. This simplifies the job and keeps the overall cost down.

The biggest problem with this type of work generally comes from the headroom that is lost by installing larger pieces of lumber. The loss is usually only a couple of inches of so, but it can be enough to create a short room. Moving existing collar ties up can often eliminate this problem. Another option can involve the use of engineered joists. These joists can be of a higher load-bearing capacity than a standard piece of lumber, even though the height of the special joist is lower. This, however, is something you will have to consult local professionals about.

BASEMENT CONVERSIONS

Basement conversions can call for some significant floor work. Most basements are equipped with a concrete floor. Many people install finish floor covering directly over the concrete, and this works fine in most cases. However, if the basement floor is forever damp, the moisture can damage finish flooring. One way to overcome this is to build a false floor over the concrete. Using pressure-treated lumber to frame a floor structure under such circumstances is a good idea. The pressure-treated wood is not affected greatly by the damp floor, and the air space between the concrete and the wood sub-flooring adds a level of protection for the finish flooring.

There are two common approaches to building a floor structure over a concrete floor. You can build a conventional floor system by attaching band boards to the basement walls and installing floor joists, on edge, as you would in conventional framing. If you need height between the concrete and wood floor to install mechanical equipment, this is a good approach to take. However, if your only goal is to install a false floor to avoid dampness and the cold that is often transferred from concrete to a finish floor, there is an easier way.

You can build a floor system out of pressure-treated 2-x-4s. Lay the lumber flat on the concrete. Attach the lumber with a powder-actuated nailing tool. This will raise the floor level by about 1½ inches. Much less than if you were to install floor joists on edge. Headroom is often a consideration in basement conversions, so the less you have to raise the floor, the better. Once the screeds are nailed into place, you have an elevated wood surface to attach subflooring to. This is an economical, and effective, way to create a false floor in a basement.

FLOOR REPAIRS

Floor repairs are sometimes necessary in remodeling jobs, especially when kitchens and bathrooms are involved. Because water often leaks through the floors of bathrooms and kitchens, damage occurs more often to the floor joists under these rooms than in any other location. I can't count the number of times I have removed old floor covering in a bathroom to find saturated, black subflooring and damaged floor joists.

If you have visual access to the floor system in the area where your work will be concentrated, you can check for damaged joists and subflooring before you ever bid a job. You should do this, because many customers will be unsympathetic if you come to them after having entering into a remodeling contract and try to explain that you hadn't figured on replacing subflooring or bad joists. Your best defense against lost profits is early detection and clear communication.

Rotten subflooring is easy enough to repair. You just cut out the bad sections and install new sections (Figures 7.7 through 7.9). Joists, however, are not so easy to replace. However, the chances are good that there is no reason to remove a damaged joist. Why bother with massive, destructive work when you don't have to? There are several ways to correct the problems caused by weakened floor joists that don't involve the removal of any existing joists. To illustrate this, let's look at a few examples.

Assume that you are remodeling a bathroom where the toilet has been leaking for years. The two floor joists on either side of the toilet are both in bad shape. What are you going to do? First, assess how far along the joists the damage runs. It probably only extends a few feet down the structural members. If this is the case, you are in luck. Simply cut out the rotted sections of each joists. Make your cuts so that both of the joists wind up to be the same

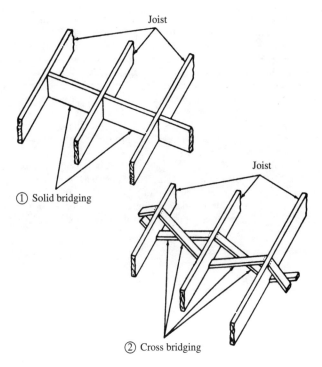

Joist

Joist

① Solid bridging

② Cross bridging

FIGURE 7.7   Types of bridging. (courtesy U.S. government)

length. Now all you have to do is install a header to attach the remaining good ends of the joists to. Position your header so that the severed joists butt up against it on the face side. The ends of the header will terminate at the neighboring joists. Ultimately, your header will span between two full-length joists, the severed joists will connect to the face side of the header, and your problem will be solved. It might be necessary to add some additional blocking between the full-length joists, depending upon how much of the bad joists you were forced to remove.

Let's study the previous example from a different direction. Let's say that the joists are damaged for most of their entire length. This makes cutting and heading them off impractical. So, what should we do? Keep in mind that you have good working conditions, because the subfloor in the bathroom has been removed. Well, suppose we install new joists right alongside of the old joists? Sure, that could work. In many cases, it is possible to slide new joists into place beside damaged ones. This is a lot easier than trying to remove old joists to install new ones. However, suppose the joist span is too long for you to get full-length joists put into place?

If you have a situation where replacement joists are too long to get into place, just cut them in half. Slide each half of the joist into place and nail it to

A. Joists attached with metal joist hangers

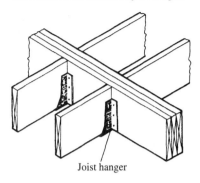

Joist hanger

B. Joists bearing on 2- by 2-inch ledger

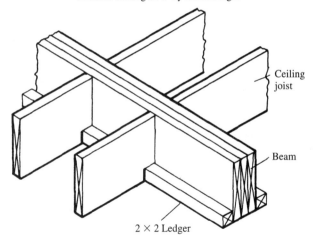

Ceiling joist

Beam

2 × 2 Ledger

FIGURE 7.8 Framing a flush ceiling beam. (courtesy U.S. government)

the damaged existing joist. Once the new lumber is where you want it, install a screw-jack post under it. You might have to create a solid base for the jack to sit on if you are working in a crawl space. A solid cinderblock will accomplish this task easily. Once the jack is set up under the joist, where the two halves meet, you've got a good sturdy support.

There is a third simple way to deal with joists where only a short section of the lumber is damaged. Let's say that you have a joist where the top section has rotted for a length of about 2 feet. What's the easy way to fix

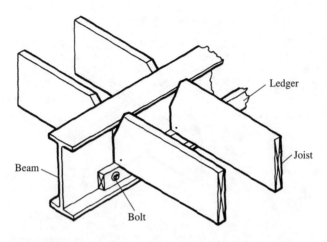

FIGURE 7.9   Steel beam used in conjunction with wood joists. (courtesy U.S. government)

this? All you have to do is scab on a new section of lumber to strengthen the rotted one.

It is advisable to install new lumber on each side of the damage joist. If the damaged section is 2 feet long, the repair sections should be perhaps 6 feet long. You must make sure the repair sections attach to the existing joist where there is adequate strength for the coupling to be made. You can nail the new sections into place, or you can bolt them to the old joists. Either way, the new sections will give plenty of strength to the weakened joist.

CHAPTER 8
FLOOR COVERINGS

Floor coverings are a big part of a home. The type of flooring used will set the tone for the residence. Remodelers are often faced with having to replace existing flooring. Builders must choice good product lines to offer their customers. Whenever floor coverings are used, they are seen for a long time. If the flooring doesn't wear well, it shows. You must be aware of how to guide your customers toward a suitable flooring, or your reputation as a builder or remodeler might be tarnished.

REPLACING FLOOR COVERINGS

Replacing floor coverings is a typical part of many remodeling jobs. There are few rooms remodeled where the flooring is not replaced. Some existing floors are left in tact and cleaned, but a majority of them are replaced. Knowing this, you should be up to speed on various flooring options (Figures 8.1 and 8.2). It is also necessary for you to know what to expect in preparation for a flooring replacement. We'll talk about types of flooring in just a little while. For now, let's concentrate on the other issues that might become a part of your next flooring job.

Carpet Replacements

Carpet replacements are probably the easiest type of flooring replacements. The very nature of carpet lends itself to hiding little flaws in existing subflooring. When a carpet pad is laid, it compensates for most minor imperfections in the base flooring. By the time carpet is installed, small ridges, little nail pops, and similar situations are not noticed. This doesn't mean that you should take new carpeting for granted.

Carpet	Qualities
Polyester	Bright colors, resists mildew and moisture, stays clean
Olefin	Very durable, resists mildew and moisture, very stain-resistant
Wool	Durable, abrasion-resistant, reasonably easy to clean, should be protected against moths, resists abrasion.
Acrylic	Resists mildew, resists insects, remains clean, resists abrasion
Nylon	Extremely durable, resists abrasion, resists mildew, resists moths, remains clean, tends to create static electricity

FIGURE 8.1 Carpet features.

Carpet	Cost
Polyester	Moderate cost
Olefin	Prices vary
Wool	Expensive
Acrylic	Moderate cost
Nylon	Prices vary
Polyester	Moderate cost

FIGURE 8.2 Cost comparison of carpet fibers.

Because carpet selections are frequently made from very small samples, it is easy for a pattern or color to appear differently when it is installed. Just the type of lighting being used when carpeting is picked out can give it a color that is not the same as it will appear once installed in a home. This is a situation that you must be aware of and that you should inform your customers of. It will prove advantageous for you to have customers sign a selection sheet before you purchase or install any flooring.

As common as it is for carpet colors to seem as if they have changed from the time they left the showroom to the time they are installed, it is even more common for patterns to give the impression of magical changes. Try to have your customers make selections from the largest samples available. The bigger the sample is, the less likely it is that the customer will be disappointed later.

It is difficult to find anything more than a small sample for customers to choose from. When this is the case, you can suggest that the customer take the sample home and look at it. Hopefully, your supplier will have enough samples to allow in-home selection. This process will, at least, give customers a view of the carpet in the same light that it will be seen in once it is installed.

When your customers pick a particular carpet for their home, record all available information from the sample on a selection sheet. Have your customers compare the information on the sample sheet with that found on the carpet sample. Once they confirm that the information on the selection sheet is accurate, have them sign in acceptance of their choice. This will not prevent the homeowners from getting angry when the carpet is installed, if it is not what they have envisioned, but it will give you solid footing to go to court on.

When you do an estimate that will require the installation of new carpeting, you have to check the existing condition of the subflooring. It will be fine in most cases. However, if you suspect problems with the base flooring, now is the time to express your concerns. Don't wait until you contracted the job to tell customers that new subflooring will be needed to make a satisfactory installation.

Who is going to assume responsibility for the old carpeting and pad that will be removed? This is normally considered to be the responsibility of the remodeler. With the increasing cost and difficulty associated with debris removal, you should determine who is going to take charge of the old material before you agree to install new floor coverings.

Seams in carpeting is another issue that you should discuss with your customers before a job is taken. Most carpet is available in a maximum width of 12 feet. There are some brands that can be purchased with widths of 14 feet, but you need to know what the maximum width of your carpet line is before you sit down with potential customers.

Making a seam with carpet is not difficult, and the seam is often hard to spot (Figure 8.3). However, seams sometimes puff up. This gives their location a noticeable hump, and it might be all it takes to turn a nice customer into a nasty one. The more you prepare a customer before work is started, the smoother the job will go.

- Carpet that is of an even height and tightly spaced with uncut loops has a hard texture that wears very well. This type of flooring works well in office environments and in other high-traffic areas.
- Yarns that are cut evenly and twisted tight are comfortably soft to walk on. The texture of this type of flooring reveals footprints, but the carpet is good for formal rooms that do not endure a lot of foot traffic.
- Rooms that take a lot of abuse can benefit from a weave that is made up of cut yarns that are twisted together upon themselves. Texture in the carpet is rough, but the material hides dirt extremely well and enjoys a long life.
- A carpet that has a pile of uneven height with patterns in the pile and uncut loops is easy to clean and stands up to excessive traffic.

FIGURE 8.3 Characteristics of carpet weaves.

Vinyl Flooring

Vinyl flooring is often used in kitchens and bathrooms. Because these two rooms are often the first to be remodeled in a home, it stands to reason that remodelers deal with a lot of vinyl flooring. It is usually sheet vinyl, but it is sometimes in the form of square tiles. In either case, vinyl is very different from carpet in terms of subfloor preparation. Carpet can be laid over a so-so subfloor with acceptable results. This is not the case with vinyl.

When you look at a job where the vinyl floor will be replaced with a new one, there are many decisions you must make before presenting a price. Will the new vinyl be installed directly over the old vinyl? How hard will the old vinyl be to remove? What condition is the subfloor in? Will underlayment have to be installed? Is the room large enough to cause problems with seams? All of these are valid questions.

It is possible to install new sheet vinyl over existing vinyl. Personally, I don't recommend it, but it can be done. The existing floor must be in good shape and it must be clean. By clean, I mean free of wax, grease, or other elements that will interfere with the holding power of adhesives. If the old floor has cuts, holes, humps, or other defects in it, you should plan on removing it. Vinyl, unlike carpet, is unforgiving. If the base upon which vinyl is installed is not flat and in good shape, the faults will show through the new flooring.

Carpet removal is pretty simple. The pad can be a pain to get out, but normally the removal work goes well. This is not always the case with vinyl. Some vinyl floors are taped into place, and they pull out easily. Sheet vinyl that was installed with a high-quality adhesive is much more difficult to get out. Vinyl tiles will usually come out if they are heated with a heat gun, but the old adhesive, from sheet vinyl or vinyl tiles, will normally have to be dealt with. Most remodelers who I know don't attempt to remove the old adhesive, they just put down new underlayment. The cost of this material is offset by a savings in labor.

Sheet vinyl that has to be installed with seams presents a liability. The seam might work loose and curl up. Heavy traffic can affect the seam and cause the flooring to peel back. If seams can be avoided, they should be. Even going to extra expense for a higher grade of flooring that comes in a wider width is worthwhile if you can eliminate seams.

New underlayment is always a good idea. The better surface you have to work with, the less likely you are to have troubles down the road. I won't take a vinyl job unless the property owner is willing to authorize the installation of new underlayment. If the existing subfloor is found to be in satisfactory shape, I credit back the costs involved with installing new underlayment. However, when underlayment is needed, I already have authorization and a budget for it. I find it is needed more often than not.

Rotted subfloors have been a consistent source of problems for me in bathroom remodeling. This is not to say that every bathroom floor has to be replaced, but a certain percentage of jobs will require all new subflooring and underlayment. I probe the subfloor of a bathroom to see if the subflooring has rotted. Common locations for problems are around the bases of toilets and the edges of bathtubs. There is occasionally a problem under a lavatory, but not very often.

Kitchen subflooring suffers from water damage sometimes. Finding a bad floor in a kitchen is more rare than finding one in a bathroom, but the odds are good that you will run into a rotted subfloor in a kitchen at some time. The dishwasher might leak or the kitchen sink might be at fault. A little probing during an estimate call is good insurance. It also builds confidence in your potential customers. If you are the only contractor to probe likely problem areas for possible water damage, the homeowners will take notice. Your attention to detail should impress them favorably.

Other Types

Other types of flooring are not replaced as often as carpet and vinyl. For example, hardwood flooring is almost never replaced. It is often refinished, but rarely replaced. Tile floors endure the test of time, and they are not normally slated for replacement. On the occasions when a long-lasting floor is going to be replaced, attention must be paid to the base flooring conditions. You might very well have to contend with the removal of a mortar base or the installation of new subflooring. Take these expenses into consideration before you offer any price estimates to customers.

FLOORING OPTIONS

Flooring options are abundant. You can offer your customers hundreds of choices in vinyl flooring. The same can be said for carpet. If the customer wants a wood floor, you can suggest a narrow board or a wide plank. There are several types of wood that you can recommend. In terms of tile, there is a host of choices awaiting an excited homeowner. Then there are less conventional flooring options, such as brick. Let's take a few moments to review some of the options that you might want to offer your customers.

Carpet

When you are discussing carpet with your customers, one of the first considerations should be the pad that will be used in conjunction with the carpet. To set a budget for carpeting, you must factor in the cost of both carpeting and padding. Some types of carpeting, like commercial-grade carpeting, have an integral pad, but most carpeting used in homes will require a separate pad. The quality of this padding can have a tremendous affect on the carpet being installed. A great carpet on a cheap pad will not look as good or last as long as a moderate carpet on a great pad. The carpet pad actually prolongs the life of carpeting. It is also responsible for having footprints bounce out of the carpeting.

Don't allow your customers to skimp on the padding. If the flooring budget is tight, recommend a high-quality pad and a mid-range carpet.

Once you get past the pad issue, you are ready to discuss types of carpeting. There are several types of carpeting to choose from, and each type has its own advantages and disadvantages.

Nylon.   Nylon carpet, for example, is an excellent all-around choice. The fibers are extremely tough and durable. Mildew and moths don't present problems for nylon, and this type of carpet is available in many price ranges. The biggest drawback to nylon is its tendency to build up static electricity.

Wool.   Wool carpeting is not used much in today's construction and remodeling market. The expense of wool is one reason why it is not installed more often. While wool is durable and resists abrasion, it is susceptible to moths. Cost is probably the biggest shortcoming of wool carpeting.

Polyester.   How does polyester stack up in the carpet industry? It ranks high in performance and popularity. The wide array of bright colors available in polyester make it eye-catching. Because it resists moisture and mildew, it is a practical choice. The price tag on polyester falls into a mid-range that most homeowners can afford. Anyone considering new carpet should give some thought to a polyester product.

Acrylic.   If your customer is looking for a carpet that seems to do it all, acrylic might be their choice. This material comes in a vast array of colors. It resists moisture and mildew, and it doesn't crush down easily. Add to this its ability to stand up against abrasions, and you've got a winner. One drawback to acrylic, however, is its likelihood of shedding. The material will sometimes produce little balls of fluff. This tendency to pill up might work against acrylic, but overall, it is a good, moderately priced carpet fiber.

Olefin.   If children are going to be active on a carpet, olefin might be the best choice. This material is super stain-resistant. It resists moisture, and it is non-absorbent. Prices for olefin cover a broad spectrum. The cheapest versions might crush under heavy traffic, and this is probably its weakest point.

Vinyl

Vinyl flooring is very popular in bathrooms and kitchens. It is also installed frequently in laundry rooms and entryways. Eat-in breakfast areas are normally floored with vinyl. Other rooms and areas of homes sometimes utilize vinyl flooring, but the locations just mentioned are, by far, the most popular places for vinyl. When it comes to vinyl flooring, sheet vinyl is the most widely used. Vinyl tiles are available, but they seldom see much use.

When you set out to shop for sheet vinyl, you could be amazed at the wide range in prices. One piece of flooring might be priced at less than $4 per yard while another is more than $30 per yard. Most builder-grade vinyl runs in the

$13 to $16 range in my area. Why is there such a spread in prices? There are numerous reasons, one of which is quality.

Cheap vinyl flooring might not be a bargain. The floor can be difficult to install. It can have little to no shine that will remain after a few cleanings. Within reason, you normally get what you pay for with vinyl flooring. I consider the lowest grade acceptable to be one that is approved for use in FHA and VA homes. If a material is FHA/VA approved, I assume it will work in an average home. However, if your customer has strong financial resources and an eye for decorating, a sizable upgrade in quality might be beneficial.

Vinyl flooring can be purchased in so many colors and patterns that it can take days for customers to figure out what they want. Just as we discussed earlier about carpeting, large selection samples make getting the right vinyl easier. Many samples for vinyl flooring are very small. This makes it nearly impossible to predict a particular pattern. Ask your supplier to provide you with samples that are large enough to make a customer's selection fun and conclusive.

What makes an expensive vinyl better than a cheap vinyl? The difference might be in how thick the flooring is or how well cushioned it is. The finish on the flooring, such as a no-wax finish, will affect price. Some no-wax finishes are better than others. A vinyl floor's ability to resist cuts and scratches can make it more valuable. A lot of shopping has to go into finding the best value in vinyl flooring.

Stick-on tiles are often installed by do-it-yourselfers. I used these tiles one time. The job was a kitchen in one of my old rental properties. This was the first and last time I ever used vinyl tiles. My experience has shown that individual vinyl tiles cause more warranty trouble for contractors than sheet vinyl does. Customers seem to get discouraged when trying to clean the many seams in floor covered with vinyl tiles.

Some customers will want these individual tiles, and it is up to you to decide if you are willing to install and stand behind them. The use of individual tiles in a foyer can be justified, but I would shy away from them in most other locations.

If you agree to install these tiles, make sure your installer does the job by the book. Insist that all installation guidelines provided by the manufacturer are followed. I believe that many of the problems associated with vinyl tiles are due to a lack of observing manufacturer's installment recommendations.

Tile

Ceramic and quarry tile make a nice floor. In the old days, the masonry base required to set tile added a lot of weight to a floor. With today's adhesives, the amount of weight associated with a tile floor is pretty much limited to the weight of just the tile. Tile floors can last a very long time, and they are easy to clean. There are some drawbacks, though. Tile floors are generally colder than other types of floor covering. They can also be very slippery when wet, and this is a disadvantage in bathrooms, kitchens, laundry rooms, and foyers, where water is often found on the floor. Tile with a raised surface for increased traction is a safer bet in these cases.

Cost can be a prohibitive factor in a tile floor. If money is tight, tile is not a good choice. Sheet vinyl is much less expensive to buy and install, and it provides a very functional floor. However, tile opens up the arena of creativity and provides some stunning floors. One complaint with tile in kitchens is the fact that fragile items dropped on a tile floor are more prone to breakage. This is true. A drinking glass or plate dropped on a tile floor is more likely to break that it would be if it were dropped on sheet vinyl or carpet.

Where should tile floors be installed? Kitchens and bathrooms are natural locations for tile. Laundry rooms are a good place for tile, and so are foyers. Sun rooms also offer a good environment for tile. Homes that are built around a solar basis often use tile as a storage mass. When and where to use tile will be up to your customers. All you have to do is be knowledgeable about the various options and the installment process involved with them.

Wood Floors

Wood floors are not installed with the same frequency that they once were. Cost is a big factor in this. However, money isn't the only reason people don't install more wood floors. The upkeep of a wood floor can be a bit of a bother (Figure 8.4). Keeping the floors sealed, waxed, and buffed requires a lot more effort than running a vacuum cleaner over a piece of wall-to-wall carpeting. The hectic schedule that most people face today leaves little time for taking care of a formal wood floor. Yet, a number of people want wood floors.

When you get into wood floors, you must learn about the various types available. Hardwood floors are the most common, but a lot of people prefer a softwood floor. Some people want narrow strips, and others want wide planks. A few people want wood squares that are installed like tile (Figure 8.5). Standard procedure usually involves unfinished wood floors, but prefinished flooring materials are available. The grade of the wood used in the flooring has a lot to do with the finished look. Homeowners seeking a rustic appearance will like rough-grade softwood floors. A customer who wants a formal floor for a dining room will probably select an oak strip floor (Figure 8.6).

- Rent a floor sander with an edger.
- Have coarse, medium, and fine sandpaper on hand.
- Start sanding with coarse sandpaper.
- Sand with the grain of the wood.
- Use edger along baseboard trim.
- Remove all stains with chlorine bleach or oxalic acid.
- Vacuum dust from floor surface.
- Apply floor finish according to manufacturer's instructions.

FIGURE 8.4 Refinishing a wood floor.

- Brick
- Finger
- Finger laid diagonally
- Herringbone
- Foursquare
- Foursquare laid diagonally

FIGURE 8.5 Patterns for parquet flooring.

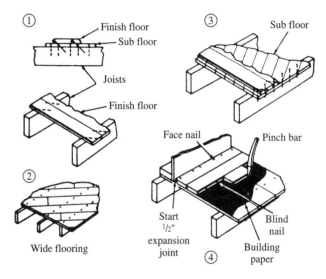

FIGURE 8.6 Methods of nailing tongue and groove flooring.

Even after a particular wood floor is decided upon, the question of a finished sealer will come up. Picking a type of finish can be as hard, if not harder, than choosing a favorite carpet color. When you are working with your customers to establish a particular finish, you should make sure that the customer sees a sample of the finish. This sample should be applied to the same type of wood that will be used on the flooring. What might look good on an oak floor could look horrible on a birch floor. It is imperative that the finish be compared on a proper type of wood.

Brick

Brick and other types of special floors are not common, but they are used from time to time. For example, a brick floor can add a lot of charm to a country kitchen. On the other hand, installing a brick floor in a powder room would normally be considered strange and unusual. Full-size bricks, when used for flooring, bring the finished floor level to a higher height. This can be overcome by using brick sections that are intended to be used as flooring.

Slate

Slate and flagstone have been used over the years as a finish flooring. A typical location for this type of heavy floor is in a foyer. The cost of installing such a floor is steep, and the floor is typically slippery when wet. Some kitchens use this type of flooring, but overall, slate and flagstone are reserved for use in breezeways and entryways.

CHAPTER 9
WALL AND CEILING SYSTEMS

In any type of project, you are likely to have to deal with framing walls, ceilings, or partitions. Framing is not particularly difficult, but doing it right isn't as easy as some people think. This chapter is going to show you various ways to deal with the framing of your walls, ceilings, and partitions.

Are you wondering what the difference is between a wall and a partition? If so, let me shed some light on the subject. Partitions are generally considered to be interior walls that separate two areas, usually two rooms. Walls are most often thought of as the exterior framing and load-bearing framing in a building. You would be correct to call either of these framing structures walls. However, the trade usually defines them as exterior walls, load-bearing walls, and interior partitions.

FRAMING EXTERIOR WALLS

Framing exterior walls can be done in a number of ways. There are many methods that can be employed to accomplish this goal, but there is one way that seems to be used by professionals on a regular basis. Most professionals build exterior walls and then stand them up and into place. This is the first procedure we will explore.

Prefab Your Walls

When you are building exterior walls, you can usually prefab them. This process allows you to frame the entire wall under comfortable circumstances. The framing is usually done on the subfloor of the structure you are building. Let me show you how this is done.

First, lay out the locations for all of your exterior walls. This is usually done by marking the subfloor with a chalk line. Once you have your layout marked, measure the length of your bottom plate. If you are dealing with a long span, you might want to build your wall in sections.

Cut your bottom and top plates to the desired length. If you are building with a standard ceiling height, you can use 2-×-4 studs that are precut to the proper height. If you are working to a unique ceiling height, cut your wall studs. Remember to allow for the thickness of your top and bottom plates. Most carpenters use one 2-×-4 as a bottom plate and install two 2-×-4s for the top plate.

Once all of your pieces are cut, you are ready to prefab your wall. Turn the bottom and top plates over onto their edges. Place your first stud at one end of the plates and nail it into place. Next, do the same with a second stud, at the other end. Many pros use air-powered nailing guns, but a regular hammer will get the job done. Once your two ends are nailed, check to make sure your framework is square. Then, proceed to install the remainder of your studs. Wall studs are typically installed with a distance of 16 inches between them, from center to center. You should use two nails in each end of your studs.

When your wall section is complete, you are ready to put it into place. If you are working with a large wall section, you might need some help in standing it up and getting it in place. Some carpenters just stand the walls up and nail them in place. Others take time to place wood blocks on the band board as a safety feature. By taking 2-×-4s, about 18 inches long, and nailing them to the band board so that they stick up past the subfloor, it gives you a bumper to butt your wall section into. This bumper makes it easier for a small crew to stand up large walls, and it reduces the risk of the wall section falling off the framing platform.

Another step to take before standing up your wall is the installation of prop braces. These prop braces are nothing more than long 2-×-4s nailed on the ends of the wall section. These braces will hold the wall up, once it is standing. You will also need some blocks of wood to nail in behind the braces where they rest on the subfloor. With large wall sections, you should prepare braces in the middle of the wall, and possibly at intervals between the center and end braces.

Once the wall is standing, position the braces to support the wall. The end of the brace should be sitting on the subfloor. Nail a block of wood to the subfloor to prevent the brace from moving. Next, nail the bottom plate to the subfloor and floor joists. That is all there is to building a prefab wall section.

Prefab Knee-walls

When you want to prefab knee-walls, the procedure is a little different. Knee-walls are meant to sit on a flat subfloor and to tie into the rafters above. This means that the top plate for a knee-wall will not be flat, like with a normal exterior wall. The top plate must be angled, to fit the pitch of the roof rafters. There are two common ways of building this type of wall on the floor.

You can cut your wall studs on an angle that will allow the top plate to be in alignment with the rafters. In doing this, the prefab procedure is about the same as described earlier. Cut your top and bottom plates. Figure the angle needed for the wall studs and cut them. Then, nail the section together. From there, you simply stand the wall up and nail it to the rafters, subfloor, and floor joists. Remember to check your wall alignment before nailing it into place permanently.

The other way of doing this doesn't use a traditional top plate. First, cut your bottom plate, and tack it into position. Then, using a plumb bob or a stud and a level, mark the bottom plate for proper stud placement. Keep in mind that for this type of framing your studs will be nailed into the sides of the rafters. Therefore, when marking your bottom plate, you must have your stud located so that it will stand next to a rafter.

When you have marked all the stud locations on the bottom plate, remove the plate from the subfloor. Cut your studs so that they will be long enough to reach from the bottom plate to a point above the rafter, allowing the stud to be nailed to the rafter.

Nail the studs to the plate, just as you would in any prefab situation. Next, stand the wall up, and get it in the desired position. Check the alignment, and nail the bottom plate in place. Then, nail the studs to the rafters, but check each stud with a level as you go. It is easy for this type of wall to get out of square while you are working with it.

When the wall is secure, cut 2-×-4 blocking to install between the studs. This blocking will be nailed horizontally between the studs, providing a nailing surface, similar to a top plate. If you don't like cutting angles, this method will work best, but the angled method is the choice of most pros.

TOE-NAIL METHODS

You could use toe-nail methods to frame your wall, but these methods are generally more difficult and not as strong. In the case of building a knee-wall, you could cut each stud on an angle to mate with the rafters. Then, you could nail the stud to the face of the rafter, rather than the side, as described earlier. On regular walls, you could erect the top and bottom plates, and toe-nail the studs between them, but I can't imagine why you would want to.

PARTITIONS

Partitions will be framed with the same basic procedures already described for walls. Partitions are typically held in place with nails driven into floor joists, ceiling joists, and connection points with adjoining walls. These connection points are frequently a place where studs have been doubled up, to allow a better nailing surface for the connection. Some carpenters use wood blocking between the studs of exterior walls to allow a nailing surface for partition walls.

WORKING WITH BASEMENT WALLS

Working with basement walls can offer additional and different challenges. The exterior walls of basements are usually made from concrete or masonry material. The floors in basements and garages are normally concrete. The walls in basements are often uneven. These walls offer little opportunity for insulation. In daylight basements, there are generally ledges that run around the entire basement, about 4 feet above the floor. This ledge is the result of the thick concrete wall giving away to the thinner wood-framed wall of the daylight section. All of these factors call for different techniques.

Attaching Walls to Concrete Floors

There are three basic ways to attach wood walls to concrete floors. You can use a drill bit to drill into the concrete. Once the hole is drilled, you can insert a lead or plastic anchor that will accept a screw. This allows you to screw the wall plate to the concrete floor. As the screw goes into the anchor, the anchor expands to hold the screw firmly. This method works, but it is very time-consuming.

Concrete nails are another way of attaching your walls to the concrete floor. Safety glasses are a must for this procedure. Nails meant for use with concrete are brittle and frequently break into pieces. When under the impact of a hammer, the nail pieces will go flying into the air. This procedure also works, but there is a better way.

With today's new tools, you can use a powder-actuated device to drive special nails through your wood and into the concrete. These tools can be rented, but smaller versions can be purchased at very reasonable prices. In the smaller version, you insert one of the special nails into the barrel of the device. Then, you put a small, rimfire powder cartridge into the chamber of the tool. You should be wearing gloves, ear protection, and eye protection when using this type of tool. The next step is to place the end of the barrel of the tool where you want the nail to be driven. Then, you hit the top of the tool with a heavy hammer or squeeze a trigger, depending on the type of tool you have, and boom, your nail is driven. These tools are great, but they can be dangerous. Read and abide by all of the manufacturer's suggestions in operating these tools and in nail and powder cartridge selection.

When you are attaching wood to concrete floors, it is best to use pressure-treated lumber. The concrete will give off moisture that can be absorbed by regular wood. Over time, this moisture might rot the wood.

Furring Basement Walls

If you are not concerned with adding heavy insulation to your existing basement walls, you can use furring strips to prepare the walls for wall coverings. It is a good idea to coat the masonry or concrete walls with a moisture sealant, before installing new walls over the existing walls.

Furring strips should be attached to the wall with an adhesive and nails, whenever possible. The same tool used to shoot nails into the floor will work on the walls.

If the basement walls are not even, you will have to place shims behind the furring strips. Use your level to determine when the furring strips are in the proper position, and then secure them to the wall.

When you want to add insulation, but don't need a lot, you can use foam boards to insulate between the furring strips. This insulation can be attached to the basement walls with an adhesive.

Building False Walls

When existing basement walls are way out of plumb or you need to add heavy insulation, building false walls is your best approach. This is done by framing walls, as mentioned earlier in the chapter, and nailing them to the floor and ceiling joists. This gives you a wall with full depth for insulation, plumbing, wiring, and heat. It also allows you to have a straight and even wall surface.

Working with Ledges

In daylight basements, it is common for a ledge to run around the perimeter of the basement. This type of wall problem can be dealt with in two ways.

You can frame a new wall between the ledge and ceiling joists to give you a straight, vertical wall. However, if you do this, any existing windows will be set deep into the wall, requiring a window box. Some people like this look and enjoy having the window box to set items in.

If you don't want the window box, you can leave the ledge and trim it out with an attractive trim board. This gives you a finished ledge that can serve to hold everything from collectibles to cocktails. Your basement can serve many needs. It can be an exercise room, an office, gameroom, family room, den, or whatever your heart desires.

CEILINGS

Ceilings are a necessary part of any finished room. The remainder of this chapter is going to show you how to work with various types of ceilings (Figures 9.1 through 9.3).

Garage Conversions and Ceilings

You are likely to run into one of two types of situations in framing garage ceilings. In one situation, you will not have any framing to do. This will be

Material	Use
Wood-fiber	Can be applied over plaster or drywall with adhesive
Mineral-fiber	Drop-in panels that work with a grid system
Fiberglass	Drop-in panels that work with a grid system

FIGURE 9.1 Ceiling tile applications.

Type of material	Cost	Features
Mineral-fiber	Very expensive	Noncombustible
Fiberglass	Mid-range price	Might be fire resistant
Wood-fiber	Inexpensive	Might be fire resistant

FIGURE 9.2 Ceiling tile comparison.

Defect	Cause
Cracked ceilings	Settlement in the building or foundation
	Vibrations in the building or foundation
Nail pops	Nails pulling loose
Drywall tape coming loose	High humidity
	Improper installation

FIGURE 9.3 Probable causes for ceiling defects.

the case when you already have ceiling joists, collar ties, or truss bands in place. If you look up in your garage and see framing that will allow you to attach to a flat surface, for a smooth ceiling, you are all set.

If you are converting the lower level of your garage, you should not need to do any additional framing for the ceiling. If you are converting the attic of your garage, you might already have collar ties that will allow the installation of your finished ceiling. If you don't have a flat surface framed up, you can simply nail collar ties across the rafters to make a flat ceiling. This procedure allows easy installation of a finished ceiling. You might use drywall, acoustical tiles, or tongue-and-groove imitation planks.

If you choose ceiling tiles or planks, installation can be easy and quick. You simply nail a metal frame-work to your joists, and hang the ceiling material. Small metal tabs hold the ceiling material in place.

Attic Conversions and Ceilings

Attic ceilings are like the ceilings in the upper level of garages. If you don't already have suitable framing for a ceiling, adding collar ties will solve your problem. Even if you choose to hang your ceiling to the rafters directly, to create a vaulted ceiling, remember your needs for proper ventilation. It is a wise idea to install collar ties high on the rafters in these cases, to allow a small flat ceiling for ventilation.

Basement Conversions and Ceilings

All basements are equipped with a ceiling structure, but it might not be to your liking. This is especially true if it is littered with pipes, wiring, and duct work, resulting in low head clearance. Let's take a look at how you can work around some of the common problems with basement ceilings.

Pipes. It is not unusual to find pipes running beneath the ceiling joists in unfinished basements. Many homeowners see these pipes and automatically assume they must install a hanging ceiling to hide the pipes. This is not always the case.

If you are willing to go to the trouble, and expense, most pipes can be relocated and raised to be hidden in the joists. There might be some drain pipes that will not allow this luxury, but most pipes can be raised. If it is not too expensive, it will be worth your while to make the necessary adjustments to allow the installation of a standard ceiling. Real estate appraisers are not kind to cheap, amateur-looking hanging ceilings. However, don't get me wrong, there are some very impressive ceiling options available if the pipes cannot be moved.

Wires. Wires are frequently stapled to the bottom of the ceiling joists in basements. These wires could be moved into the joists, but there is an easier way to hide them. You can simply drop the ceiling joists down with furring strips. The furring strips will allow a chase-way for the wires, while allowing the installation of a traditional ceiling.

Duct Work. Duct work that hangs from basement ceiling joists is not so easy to hide. However, there are ways to improve the look of these metal monsters. If you have duct work, you will have one large trunk line that runs most of the length of your basement. This large duct work must stay below the joists and be boxed in.

The smaller supply and return ducts can usually fit in the space between joists. It might take a little cutting and metal bending, and bleeding, but you can move most of these small ducts. I'm serious about the bleeding, even seasoned professionals often cut themselves on the sharp edges of duct work; be careful.

The old holes in the trunk line, from the relocated ducts, can be covered with new sheet metal that is held in place with screws. Before you attempt such a conversion with a heating system, consult with a local, licensed expert to be sure you will not harm your customer's heating system.

Beams.   Beams are a frequent fact of life with basement conversions. They are unwanted, but usually needed. One way to make the most of your beam is to box it in and install recessed lighting in the box, so the box has a purpose, other than just hiding a beam.

You can build your false box around the beam and attach it to the ceiling joists. This allows you to make it any reasonable size that you like. If you prefer to simply wrap the beam, drill holes in the beam with a high-quality drill bit. Wear eye protection, and be aware that the metal shavings will be hot. Bolt 2-×-4s to the beam and create a box just slightly larger than the beam. Then you can attach drywall to the wood and cover the beam with a close-fitting box.

Standard Ceilings

Standard ceilings are made as the result of either floor joist for a room above, or as a part of a roof structure. We will talk about roof systems in chapter 12, so you will get more information there on traditional, new-installation ceilings.

CHAPTER 10
EXTERIOR WALL FINISHES AND TRIM

Siding and exterior trim is a field where a contractor can set up shop as a specialist. Many do, and these specialists are hard to compete against if you are a general remodeler. It stands to reason that, if you do the same type of work day in and day out, you should get very good at it. This is usually the case with siding specialists. Because they are set up only for siding and exterior trim and that is the only work they do, they are fast. Being fast translates into either increased profits or lower bids, depending upon the approach one chooses to take. This is why it is tough for a general practitioner to compete with any type of specialist.

SHOULD YOU SPECIALIZE?

Should you specialize in siding installations? Maybe. If you like siding work, live in an area with a decent amount of aging homes, are set up for siding, and can find fast mechanics to work with you, siding is a viable specialization. If any of these conditions are not met, you might not do so well in a siding business.

Siding is a very competitive field. Because licensing requirements in this area of work are typically lax, there is a lot of competition. Much of it is easy to overcome, but then you have those dedicated crews that work from dawn to dusk. Siding can generate a huge profit, in terms of percentages, but the actual cash value of a siding job is nothing to compare with an attic conversion, a room addition, or building a new house. Are you willing to give up the big scores to get a lot of little jobs? Depending upon your personal situation, you might actually be better off with a lot of little jobs. There are more of them, but you don't have to be accountable for as many people on small jobs. A siding crew of just three people can be very productive. Compared to the number of people involved in a major conversion or building project, three people are much easier to keep tabs on.

SIDING IS SO SIMPLE

Some contractors believe siding is so simple to install that there is nothing major that can go wrong during this phase of construction. I say that this line of thinking is nonsense. True, siding is not as complicated or as perilous as removing a complete roof structure and building a second-story addition. However, there is still plenty that can go wrong on a siding job. In fact, let's discuss some of the problems that might crop up on a siding job.

Pump Jacks

When you or your crews are installing siding, do you use pump jacks or staging? One or the other of these types of elevated platforms is needed, and either one of them can damage a lawn. How do you protect a person's grass when you are setting up for a remodeling job? Do you put blocks of wood under your staging to level it? How often do you use wide boards under your equipment to distribute weight more evenly? You might not have run into a really picky customer yet; however, if you stay in the remodeling business long enough, you are bound to encounter some customers who will make your life miserable. The catalyst to fire up this nasty customer could be the marks your scaffolding leaves in a lawn.

How can you hedge your odds against upsetting a customer with depressions in the lawn? Putting wide boards or plywood under the supports of your work platform will help. At the least, they will prevent deep holes from being punched into the earth. However, the boards might kill any grass that is beneath them. As long as you don't keep the boards in one place for more than a day, the grass should recover. However, if you set up and leave your equipment in the same place for several days, the customer's lawn might turn brown in the spots where you had your boards. Because there is a level of risk here, you should talk to your customers before doing any work and describe the potential problems associated with installing new siding. This little chat won't stop bad things from happening, but it can make your customers more forgiving.

Broken Windows

Broken windows are more common than you might think on siding jobs. A cross bar from a section of staging can swing loose, when being disassembled, and break the glass out of a window. Careless workers sometimes ram cross bars, jack poles, and even siding into windows. There is also the occasional hammer that is dropped onto the work platform, which then bounces into a window and breaks it. Accidents happen, but they seem to happen much more frequently with some contractors than with others. You can't really prepare a customer to accept a broken window, because there is no legitimate excuse for such a thing to happen.

Inside Breakage

Inside breakage of personal affects is common on siding jobs. Too many contractors fail to tell their customers what to expect when old siding is removed and new siding is installed. The banging done on exterior walls is enough to make picture frames fall. There is also plenty of force to make knick-knack shelves drop to the floor. A broken picture frame might not seem like a big deal to you, but it can be to a customer. Let me tell you a short story of such a case.

I met an elderly lady last year when she wanted some work done in her home. As I walked through the house, I noticed an extensive collection of decorative plates. The kind sold in gift shops. The plates were from all over the country. In addition to the plates, there were numerous empty holders on the shelves, and a few of the plates had obviously been glued together after having broken. As small talk, I inquired about the background of all the plates.

The lady told me that she and her late husband had traveled all over the country, and they had collected plates as they went along. With her husband being recently deceased, the plates were all she had left to remind her of those happy times out on the road. The way this woman talked about those plates, you would think they were her grandchildren. It was very obvious that each and every plate on those shelves was dear to her.

As I learned the history of each plate, one by one, I found out why there were so many empty holders on the shelves. I also learned how the damaged plates that remained in the collection had come to be broken. A siding contractor had been hired to replace the lady's existing siding with vinyl siding. During the work, the contractor's crew had created enough disturbance on the outside of the home that the missing plates had been knocked out of their resting places. A good number of the plates were destroyed beyond repair. Needless to say, the lady was heartbroken to see pieces of her best memories go into the trash. Even if the contractor had been remorseful, which she said he wasn't, there would have been no way to undue the damage done.

If the contractor had advised this lady of the potential for the plates being broken, I'm sure she would have removed them from their shelves until all of the work was done. The contractor's negligence cost the lady a lot, in terms of emotional loss. He also didn't do himself any favors in getting new job referrals. You can bet the homeowner would not be quick to recommend such a contractor to friends.

If you've been doing siding work for awhile, you know what's likely to happen on most jobs. Tell your customers what to expect. I've heard contractors say that they don't inform their customers of what might happen because they are afraid of losing the job. I doubt that you will ever lose a job by being honest and helpful with your customers; however, if you do, you probably wouldn't have wanted the job anyway.

Bats, Rats, and Things

When you are removing siding from a house, you might discover bats, rats, and all sorts of other things. The exterior walls of houses can be home to a multitude of wild creatures. Snakes, squirrels, bats, mice, bees, and other living creatures often take up shelter in the walls of homes. Pulling a piece of siding off of a high gable and coming face to face with startled creatures can make you a little weak in the knees. If a swarm of angry hornets come flying out fast to see who just destroyed the side of their nest, which was attached to the back of a piece of siding, a siding installer can be given quite a scare. Being perched some 20 feet above the ground on a small walk board is not a good place to be when a bunch of mean hornets are looking for something to attack.

Depending upon where you work, you might never run into any of the wild things that I've just told you about. However, I assure you it can happen, because it has happened to me. I grew up in the country, and when I started working in the trades, many of the jobs were associated with old houses. Many of these houses were out on some acreage, and it was not at all uncommon to find all sorts of living things in the homes. I've been nose to nose with snakes, and I've been dropped low by fleeing bats. Bees have attacked me, and rats, big Norway rats, have given me the creeps. Squirrels have scampered by me, and raccoons have stared out of walls and ceilings at me. Believe me when I tell you that almost anything can be waiting for you on the other side of that siding you are removing. Of course, more modern homes don't offer the same risks, because they have more consistent sheathing between wall cavities, attics, and siding.

If you are working with exterior trim and have to remove a section of soffit, be careful. This is a favorite hang-out for bees. It is also a place where snakes might seek refuge. Can you imagine yourself pulling down a section of soffit only to have a large, lively snake fall into your face? It could happen, so be careful.

Nails

I would like to assume that all siding installers know what types of nails to use, but I know this just isn't the case. How many houses have you seen where siding was installed with the wrong nails and the nails caused rust streaks to run down the siding? I've seen several over the years. There is no excuse for this type of problem. The right nails cost more than the wrong nails, but the difference is a small price to pay in order to preserve the siding. I always insist on stainless-steel nails when working with wood siding, and I hope that you are knowledgeable and responsible enough to use the proper nails on your jobs (Figures 10.1 through 10.3).

Penny size "d"	Length	Approximate number per pound common	Approximate number per pound box	Approximate number per pound finish
2	1"	875	1000	1300
3	1¼"	575	650	850
4	1½"	315	450	600
5	1¾"	265	400	500
6	2"	190	225	300
7	2¼"	160		
8	2½"	105	140	200
9	2¾"	90		
10	3"	70	90	120
12	3¼"	60	85	110
16	3½"	50	70	90
20	4"	30	50	60
30	4½"	25		
40	5"	20		
50	5½"	15		
60	6"	10		

Note: Aluminum and c. c. nails are slightly smaller than other nails of the same penny size.

FIGURE 10.1 Nail sizes and number per pound.

Electrical Wires

It should go without saying when talking to professional remodelers, but I'm inclined to mention power lines. Every siding job will involve some work around power lines. If a home is served by overhead power, the siding around the weatherhead can make working conditions dangerous. Respect electricity and the damage it can do. Before you set up near an electrical service, have the utility company come out and wrap the lines to reduce your risk of injury or death.

TYPES OF SIDING

What types of siding do you offer your customers? Are you a full-line installer, or do you only work with vinyl? Is wood siding the only type that you

Length	Gauge numbers
¼"	0 to 3
⅜"	2 to 7
½"	2 to 8
⅝"	3 to 10
¾"	4 to 11
⅞"	6 to 12
1"	6 to 14
1¼"	7 to 16
1½"	6 yo 18
1¾"	8 to 20
2"	8 to 20
2¼"	9 to 20
2½"	12 to 20
2¾"	14 to 20
3"	16 to 20
3½"	18 to 20
4"	18 to 20

FIGURE 10.2 Screw lengths and available gauge numbers.

will install? Are you willing to install any type of siding a customer requests? There are only a few types of siding that make up most of the market; however, in addition to these standards, there are other types of siding available (Figures 10.4 and 10.5). You might get requests for any type of siding.

Cedar Siding

Cedar siding is very popular, especially in new construction. The cost of cedar is one of its few disadvantages to a homeowner. Cedar looks good, it can be used on almost any type of house with desirable results, and it lasts for a long time. However, it should be sealed or stained promptly to prevent discoloring. I've seen a lot of homes where cedar siding was not treated promptly, and the siding turned gray. Some people like this weathered look, but don't count on your customers liking it. You should advise your customers that cedar will discolor if it is not sealed within a reasonable time. Then, if the siding discolors, the customer can't plead ignorance.

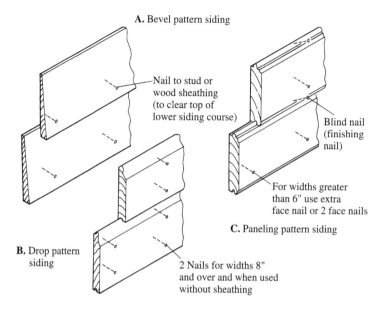

A. Bevel pattern siding

Nail to stud or
wood sheathing
(to clear top of
lower siding course)

Blind nail
(finishing
nail)

For widths greater
than 6" use extra
face nail or 2 face nails

C. Paneling pattern siding

B. Drop pattern
siding

2 Nails for widths 8"
and over and when used
without sheathing

FIGURE 10.3   Methods of nailing wood siding. (courtesy U.S. government)

Material	Care	Life, yr	Cost
Aluminum	None	30	Medium
Hardboard	Paint Stain	30	Low
Horizontal wood	Paint Stain None	50+	Medium to high
Plywood	Paint Stain	20	Low
Shingles	Stain None	50+	High
Stucco	None	50+	Low to medium
Vertical wood	Paint Stain None	50+	Medium
Vinyl	None	30	Low

FIGURE 10.4   Siding compared.

Material	Advantages	Disadvantages
Aluminum	Ease of installation over existing sidings Fire resistant	Susceptibility to denting, ratting in wind
Hardboard	Low cost Fast installation	Susceptibility to moisture in some
Horizontal wood	Good looks if of high quality	Slow installation Moisture/paint problems
Plywood	Low cost Fast installation	Short life Susceptibility to moisture in some
Shingles	Good looks Long life Low maintenance	Slow installation
Stucco	Long life Good looks in SW Low maintenance	Susceptibility to moisture
Vertical wood	Fast installation	Barn look if not of highest quality Moisture/paint problems
Vinyl	Low cost Ease of installation over existing siding	Fading of bright colors No fire resistance

FIGURE 10.5   Siding advantages and disadvantages.

Pine Siding

Pine siding is one of my favorites. It is easy to work with, it's wood, it can be stained or painted, and it gives a good finished appearance. Pine is substantially less expensive than cedar, and once it is installed and stained or painted, it is hard to tell that it is not cedar. Some people don't like the knots found in most pine siding, but I think they add to the character of a home.

I've used pine siding for many years, and I've never been disappointed with it. I do make a point of having it stained, or painted, quickly. Pine siding that is left in its raw condition and exposed to rain will turn black. It doesn't take long for the discoloration to set in.

I recently discovered a supplier who has a machine that stains the siding before it is delivered to a job site. When I built my new home, I gave this prestained siding a chance. It worked out great. The fee charged by the supplier to stain the siding was less than what my painters were quoting for the work, and the siding arrived in good shape. Because it was already stained, there was no opportunity for the wood to discolor. By using this procedure, I saved a little money on my staining costs, and I saved a good bit of time on the job. If you have not experienced this type of arrangement, you should check with your local suppliers to see if they can offer the same service.

Aluminum Siding

Aluminum siding is not used very much anymore. There was a time when it was very popular, but that time has passed. Vinyl siding has, in my opinion, made aluminum siding a thing of the past.

Vinyl Siding

Vinyl siding gets mixed reviews, depending upon who you are talking to. Some people swear by it, and others swear at it. I like some of the attributes of vinyl siding; however, on the whole, I am not thrilled by it. My experience has shown that real estate appraisers are not too keen on vinyl siding either. This is a big factor to a builder who is working to get the most appraised value possible for every dollar spent. However, if you are not concerned with a maximum appraisal and you are installing what a customer wants, vinyl is fine.

One of the biggest selling points associated with vinyl siding is that it is said to be maintenance free. While vinyl siding never needs painting, it sometimes does require a power washing. Damp locations can cause mold and mildew to form on vinyl siding. When this happens, a power washing is needed to clean the siding. This is less expensive than painting, but it is a form of maintenance.

Overall, vinyl offers a lot to homeowners. The fact that the color is a part of the vinyl is a big advantage. Because the color won't chip or peel, it lasts for a very long time, and it never needs to be painted. A house that is equipped with vinyl-clad windows, vinyl siding, and wrapped trim is about as maintenance free as it can get. Another advantage to vinyl siding is that it can often be installed over existing siding, reducing the expense incurred by a property owner. All of these factors combine to make an attractive sales package.

Other Types of Siding

Other types of siding exist and are used, but the ones we have just covered are, by far, used most often. Some customers might want another type of siding, such as T-1-11 (Figure 10.6), but these cases will be rare. Homeowners with expensive taste might seek a board-and-batten siding, but again, there will not be a lot of call for this. Cedar shakes might account for a very small percentage of siding requests. If you are prepared to work with cedar, pine, and vinyl, you should be in the hunt for the majority of siding work to come your way on a residential level.

Type	Use
Softwood veneer	Cross laminated plies or veneers—Sheathing, general construction and industrial use, etc.
Hardwood veneer	Cross laminated plies with hardwood face and back veneer—Furniture and cabinet work, etc.
Lumbercore plywood	Two face veneers and 2 crossband plies with an inner core of lumber strips—Desk and table tops, etc.
Medium-density overlay (MDO)	Exterior plywood with resin and fiber veneer—Signs, soffits, etc.
High-density overlay (HDO)	Tougher than MDO—Concrete forms, workbench tops, etc.
Plywood siding	T-111 and other textures used as one step sheathing and siding where codes allow.

FIGURE 10.6 Types and uses of plywood.

ASBESTOS

Siding that contains asbestos can present some health and financial problems for you. If you are going to remove this type of siding from a home, you will most likely have to hire contractors who are licensed to do asbestos abatement work. This can get very expensive, so you cannot afford to overlook this potential cost. This is another situation where a clause in your contract can save you a lot of money. The clause should discuss what will happen if the siding is found to contain asbestos. If the homeowner is willing, you can check the siding before you commit to a price to remove it. If it contains asbestos, you can make financial arrangements for its removal before you enter into an agreement.

BEFORE YOU SIGN

Before you sign a contract to install siding on an existing home, you should do a complete inspection of the property to be sided. You might discover that the walls are way out of plumb and that some compromises must be made during the installation. It is much better to discover this before the job is started. The same goes for an asbestos situation.

As with any type of work, you should sit down with your customers and discuss the job thoroughly before you sign a formal proposal or contract. There will be many details that must be worked out. For example, how will the old siding be removed from the job? Will this be your responsibility or that of the homeowner? Getting rid of a large load of siding can get expensive, so you should establish if this cost will be in your bid or not. If you just lump the removal costs into your price, without advising the customer of the expense, you might be presenting a proposal that is much more expensive than those of your competitors who did not address the issue of removal. As long as you educate your customers in the expenses to be expected, they can compare bids fairly. If they are not sure if your competitors are assuming responsibility for removal work, a few quick phone calls to the other contractors will provide the answer. These calls should be made by the potential customer.

Siding a home is fairly simple work. It is not the trade-related work that you have to worry about; it is the risk of unexpected problems that should concern you. These problems can be avoided with proper planning, a thorough inspection, and clear communication with your customers.

CHAPTER 11
INTERIOR WALL FINISHES

When we choose to talk about wall coverings, the potential topics cover a broad spectrum. We could start with paint, because it is one of the most used wall coverings. However, is it? Is paint a wall covering or is drywall a wall covering? Depending upon the phase of your job, either one could be the wall covering at issue. Once walls are framed and all rough-in work is done, wall coverings are needed. This class of wall coverings will typically include drywall and paneling, although other types could be included. If we assume drywall is used to cover stud walls, as it usually is, we can then focus on paint as a wall covering. However in addition to paint, we could discuss tile and wallpaper. Well, this chapter is going to cover both types of wall coverings and all the major options within each classification.

Contractors deal with wall coverings in a number of ways. They install them, cut into them, and repair them. Wall coverings are removed, replaced, cleaned, and altered. All of this work adds up to a need for knowledge. A good contractor must be able to offer suggestions on what the most appropriate wall coverings for a specific project are. Can you do this? If you've been in the business for a few years, you should have a good, basic understanding of the most common wall coverings. However, are you familiar with various types of tub surrounds to offer a customer when a tile wall around a bathtub needs to be replaced? Have you ever suggested that a customer use weathered barn boards as a wall covering when creating a country kitchen? There are a lot of creative angles to wall coverings that many contractors seldom think about.

When was the last time that you recommended that a customer use a brick facade on a family room wall? Have you ever offered customers an option for murals in a child's bedroom. Most contractors wait for their customers to say what they want, and then that is what the remodelers base their bids on. This is not wrong; however, if you operate in this manner, you are probably losing some jobs that you could win.

Customers tend to like contractors who make creative suggestions. If you impress a customer with your depth of knowledge and your willingness to make their home special, you've got a leg up on the competition. Many contractors are always in a rush. They feel they don't have time to tinker around with the whims of homeowners. These contractors work up an estimate, call or mail it in, and wait. Sometimes they wait for a very long time, because more creative contractors are awarded the job.

You don't have to be an interior decorator to offer helpful advice to confused customers. Put yourself in the customers' place. An average person has no idea of all the options available for various types of work. They are not professional contractors, and most of their information has come from reading magazines or talking to friends. If you are in touch with current trends, you can sweep homeowners off their feet. Price will not matter nearly as much to a homeowner who feels they have found a competent contractor who has their best interest at heart.

If you walk into a kitchen that a customer wants remodeled and assume that drywall will be used for the walls, you are probably right on target with every other contractor who comes along to bid the work. However, what happens if you suggest tile accents, stenciling around the top of the wall, or a herringbone pattern of tongue-and-groove planks. All of a sudden, you are different. You've captured the customer's attention and set yourself apart from the crowd of contractors competing for the work. This is how enterprising contractors get ahead and stay ahead.

You might feel that wall coverings is a mundane subject. Your personal feeling might be that painted drywall is cost-effective and good enough for any house. It's fine to have your own opinion, but don't shove this line of thinking onto your customers. Present them with creative options and allow them to make their own choices. I think you will find yourself winning more bids and working more consistently.

COVERING THE STUDS

What type of material are you going to suggest for covering the studs in rough framing? Drywall is the obvious answer, and it is usually the best answer. However, there are times when plain old drywall isn't best. Wood paneling has been popular off and on, and it has many good features. Would you suggest the installation of paneling in a family room? How about in a kitchen? Would you use it as wainscotting in a dining room? Does paneling have a place in a bathroom? Paneling can be used in any room, but it is not always an ideal choice.

Once you get past paneling and drywall, the options for covering studs in modern construction shrink. You could use plaster, but this is hardly ever done in modern building. Other options fall into what I consider a special-use category. Brick might be used in a family room, and so might stone. Old barn boards might make a very comfy country kitchen. Wood siding can give a rustic look to an office or study. Any of these materials can be used to cover rough framing, but they should be used in moderation and with special consideration.

Drywall

Drywall is far and away the most popular form of wall covering when it comes to closing in rough framing on the interior of a home (Figure 11.1). This material is relatively inexpensive, fairly easy to work with, durable under average conditions, and well accepted within the industry. The popularity of drywall is so great that many people never consider alternative options. As a contractor, you can go with the flow and use drywall on almost all of your jobs. However, you've already seen in this chapter how being a little different in your approach can give you a competitive edge.

There are two basic scenarios that most contractors face. They are either working with existing wall coverings or starting from scratch with bare studs. This can make a difference in a decision of whether or not to use drywall. If you are doing a facelift on a kitchen that already has drywall installed, you might choose to patch and paint the wallboard. This, of course, is affordable, fast, and pretty easy. If drywall is already in place and it is in good shape, there is little reason to tear it down, unless a customer wants a completely new look. Even then, you might be able to use the drywall as backing and apply a new type of wall covering, such as paneling, over it.

Drywall is easy to repair. If you need to open up a stud bay to allow the installation of a plumbing vent or electrical wires, you can cut into drywall with little fear of not being able to provide an acceptable patch. This can't be said for some other types of walls. All in all, drywall (Figures 11.2 through 11.5) is a fantastic wall covering.

If you rip out a room down to bare studs, or if you are building a new room, customers might want to consider something other than the day-to-day standby of drywall. If you have bare studs to work with, you are unlimited in the types of wall coverings you can offer a customer. When cost is a concern, drywall will usually prove to be the most acceptable type of wall covering. However, you must factor in all the expenses associated with it. The drywall has to be purchased. Then it must be installed and finished. Once the drywall is finished, some type of finish wall covering, like paint or wallpaper, must be

Type	Thicknesses	Sizes	Uses
Regular	¼", ⅜", ½"	4-×-6 to 4-×-14	Interior walls and ceilings
Moisture resistant	½", ⅝"	4-×-6 to 4-×-16	Base for tile in bath, etc.
Fire resistance type X	½", ⅝"	4-×-6 to 4-×-16	Fiberglass and additives in core for fire hazards or high heat areas.

FIGURE 11.1 Types of drywall or gypsum board.

Type of material	Edge types	Thickness	Length
Regular	Tapered	¼"	6–12 feet
Regular	Square	⅜"	6–16 feet
Regular	Rounded	½"	6–16 feet
Regular	Rounded	⅝"	6–12 feet

FIGURE 11.2   Specifications for regular drywall.

Type of material	Edge types	Thickness	Length
X	Tapered	½"	6–16 feet
X	Rounded	⅝"	6–16 feet
X	Square	⅝"	6–16 feet

FIGURE 11.3   Specifications for fire-resistant drywall.

Type of material	Edge types	Thickness	Length
W	Tapered	½"	8, 11, or 12 feet
W	Tapered	⅝"	8, 11, or 12 feet

FIGURE 11.4   Specifications for water-resistant drywall.

Number of sheets	Pounds of screws (1¼")	Joint compound	Tape
10	1	2 gallons	120 feet
16	1½	4 gallons	200 feet
20	2	4 gallons	240 feet
24	2½	5 gallons	300 feet

FIGURE 11.5   Quick estimating chart for drywall.

added. If you total up the cost for each of these phases of work, installing a good grade of prefinished wood paneling might prove to be cheaper. It will certainly be faster. All of this must be considered when planning on what wall covering to use.

When money is not as much of an issue as appearance, some of the more creative wall coverings might be more desirable. These could range from brick, to weathered boards, to half logs, to wood siding or planks. Special wall coverings should not be used in just any room or in just any way. However, used judiciously, these accent coverings can set a tone for a room. This is especially true when a rustic feeling is being sought, as might be the case in a country kitchen, a family room, or a study.

Problems associated with drywall are many. First, there is the dust from sanding that must be controlled. If a contractor is not careful, the sanding of drywall in a work area can invade other parts of a home, creating a considerable mess. Who cleans this mess up? You guessed it, the contractor. Time is a big drawback to drywall. The boards can be hung quickly enough, but the finishing, priming, and painting process can take weeks. Compared to prefinished paneling, where the job is done as soon as the sheets are installed, drywall is a loser in terms of time.

Another potential drywall problem can arise after a job is done. Seams can show, joints can crack, and the boards themselves can be damaged easily. This is not the case with paneling and some other types of wall coverings. Because no contractor likes call-backs, this reason could be all the proof needed to look for an alternative wall covering.

Drywall requires some type of finish wall covering. If we assume this will be paint, someone will have to maintain the paint over a course of years. The walls will become dirty, and the paint will need to be replaced periodically. A cost-conscious homeowner might be willing to pay more up front for a maintenance-free wall covering, such a prefinished paneling. This is something else that you should take into consideration when giving advice to your customers.

Now that we have beaten down drywall, let's pick it back up. The initial cost of installing drywall is usually less expensive than comparable wall coverings. Under normal conditions, drywall will last for years and years. Due to its nature, drywall is easy to repair if it becomes damaged. The decorating opportunities when working with drywall are practically unlimited. The wallboard can be covered with paint, tile, wallpaper, or other creative options. This allows a homeowner to create a custom look in a house. If light walls are wanted, they can be had. When a darker color is desirable, it is no problem to obtain. A pink room that was used as a nursery for a first child can be changed to a blue room easily for a second child. Drywall does offer a host of advantages, and it is the mainstay within the building and remodeling industry.

Paneling

Paneling has been popular off and on for years in residential construction (Figure 11.6). Many do-it-yourselfers flock to paneling because of the ease

Did you know that paneling is sold in a variety of sizes? Many contractors believe that 4-×-8 sheets are the only size of paneling available. In reality, paneling comes with widths of 30 inches, 36 inches, and 42 inches, in addition to the standard width of 48 inches. The length of paneling is 8 feet as a standard, but other lengths are available:

- 5 feet
- 6 feet
- 7 feet
- 9 feet
- 10 feet
- 12 feet

Larger sizes are available through special orders.

FIGURE 11.6 Facts about drywall.

with which it can be installed. You won't find many houses where every interior wall is covered with paneling, but it is not uncommon to discover certain rooms where paneling has been used in homes. Is paneling a good alternative to drywall? It depends on several factors.

If a customer wants wood paneling installed, then wood paneling should be installed. Should you make professional recommendations for paneling? There are times, I believe, when you should. If a customer is asking for a formal dining room or living room where wainscotting will be installed, high-quality wood paneling is a sensible solution. Paneling is also a reasonable alternative in rooms that will receive rough treatment along the walls.

If a customer calls you to build a play room for a child, what are you going to recommend for the wall covering? Concrete might be the most appropriate choice in some ways, but it will not likely be one of your options. Most contractors would recommend either drywall or paneling.

I would favor paneling in many ways. When a tricycle runs into drywall, some damage to wall is likely to occur. The same collision with a paneled wall would probably never be evident. Kids often play rough, and drywall is easy to dent and damage. Paneling is not. One drawback to paneling in a playroom is that it could become damaged to a point where replacement is needed. An example of this could be an enthusiastic painter who decided to paint a pin-the-tale-on-the-donkey game on the paneling. If this were done on drywall, the artistic creation could be painted over. With most types of paneling, the affected paneling would have to be replaced.

People frequently install paneling in basement conversions. Basements tend to be dark to begin with, and installing a dark paneling worsens the situation. Room should be light and airy, for the most part. If paneling is used, the color of the paneling should be selected carefully.

I'm sure you've seen the cheap paneling advertised in the weekend paper. So has just about every homeowner considering a remodeling project. This super-cheap paneling is no bargain. I've used it before, in some of my personal rental property, and it is not the type of product that I would install for paying customers. Perhaps I should qualify my statement. The paneling I'm discussing might work very well if it were installed over existing drywall, but when nailed directly to stud walls, the panels warp, twist, and do all sorts of crazy things. If you are going to introduce your customers to paneling, make sure they shop for quality products.

Special Wall Coverings

When you venture into special wall coverings, you must understand why they are special. Installing weathered barn boards in a master bedroom or a living room is probably not a good idea. However, putting them in a country kitchen could win you some awards. An interior brick wall is not common, but it can add warmth to a room, in terms of atmosphere, and it provides a nearly indestructible barrier for children to play around. My parents had a brick wall installed in the family room of one of their houses when I was a child, and I'm still enamored by it.

As a contractor, I've used brick in family rooms where a masonry fireplace was centered on a wall. Brick has been used in several of the kitchen jobs I've done. The use of brick gets expensive, and it can be too overbearing for a room, so don't get carried away with it.

Exterior siding, such as cedar or pine, can be used to give a room a rustic decor. I've done this in various jobs, and all of the customers have loved their new rooms. Tongue-and-groove boards also work well in this way. The use of wood siding and boards can make a very dramatic statement, and the designs that are possible when working with wood can give a home a custom, signature look.

How many bathrooms have you remodeled that were covered in water-resistant paneling? I've dealt with more of them than I care to remember. During my career, I've torn out a lot of this bathroom board, but I've only installed it once, and I would never do it again. In fact, a few months ago a customer asked me to install some of this hardboard in his bathroom. I agreed to do all of his remodeling work, except for the walls. The customer insisted on using the hardboard, and I stood firm on refusing to install it. The homeowner and one of his friends installed the wall covering themselves, and it didn't turn out very well. I tried to tell him, but he just wouldn't listen. I'm sure there must be occasions when this material is suitable, but I can't think of one.

Tub surrounds are often installed by remodelers. Most of these units are meant to be installed over water-resistant drywall, but a few are made to install directly over stud walls. In my experience, the only tub surrounds that work well when attached directly to wall studs are heavy fiberglass models. Thin fiberglass units, plastic surrounds, and other types of surrounds do best when installed over drywall.

Once you get the stud walls covered, you have to apply a finish coat in many cases. Because most walls are covered with drywall, the options for finish coats are numerous. Even when wood siding or planks are used as interior walls, some type of finish coat is needed. Brick and prefinished paneling don't call for this additional work, but most walls do. So, let's talk now about finish wall coverings.

FINISH WALL COVERINGS

Finish wall coverings come in many types, shapes, and colors. The ways these products are installed are as diverse as the products themselves. Considering the number of potential options, it is easy to understand why customers can become confused when trying to sort through the maze of choices. Good remodelers are aware of this, and they strive to clear the confusion for their customers. To better prepare you for this task, let's talk about specific types of finish wall coverings.

Priming

Priming surfaces prior to painting should be considered essential (Figures 11.7 and 11.8). A quality primer will allow paint to cover better. Even if a wall is going to be covered with wallpaper, some type of primer should be installed first. If wallpaper is applied to an unprimed wall, removing the wallpaper at a later date will be difficult. The adhesive used with the wallpaper will most likely pull off the paper coating of drywall when the wallpaper is removed. Priming the drywall prior to installing wallpaper will prevent this.

- Make sure all wall surfaces are clean and dry.
- Apply two coats of latex paint as an undercoater.
- Apply one finish coat of latex paint.
- Apply two coats of enamel undercoater to trim.
- Apply one coat of enamel finisher to trim.
- Use gloss paint in bathrooms and kitchens.

FIGURE 11.7   Tips for painting new interior walls.

- Wash walls and ceiling thoroughly.
- Fill cracks with joint compound.
- Sand and dust compound areas.
- Paint ceiling first.
- Paint walls and cut in around ceiling a little at a time.

FIGURE 11.8   Tips for painting older interior walls.

A good primer is every bit as important as a good paint. If you will be painting a wall, prime it first. It is usually a good idea to have the primer tinted to a color that will work well with the finish coats of paint used. Regardless of what you will use as a finish wall covering, buy and install a primer that is recommended for use with the finish product that will be used.

Latex Paint

Latex paint is the most common type of paint used in modern construction. This paint doesn't emit offensive odors, it cleans up easily, and it dries quickly. These are all good characteristics. Latex paint is very durable, and it resists mildew. With so much going for it, there is no wonder that latex paint is so popular. However, if you will be painting over existing oil-based paint, latex paint might not be the best choice for the job.

Oil-based Paint

Oil-based paint gives off odors that are offensive to many people. Some people have allergic reactions to the vapors. Oil-based paint is slow to dry, and cleaning up this paint is both messy and time consuming. Many professional painters believe the durability of an oil-based paint makes the negative points of it acceptable, but others disagree.

Acrylic Paint

Acrylic paint is in the latex family. This paint can be thinned with water, and it dries even faster than standard latex paint. Another favorable quality of this paint is that it covers well on almost any type of building material.

Alkyd Paint

Alkyd paint is normally used in conjunction with oil-based paint. This is a synthetic-resin paint that is thinned with a solvent. Alkyd paint dries more quickly than oil-based paint, but much slower than latex. When you have to cover existing oil-based paint, alkyd paint will cover very well.

Wallpaper

If you get involved with wallpaper, you will soon find that there are a multitude of options available (Fig. 11-9). Matching wallpaper to your specific needs is not always easy. Professionals in retail stores can be of a lot of help in this matter. However, you won't have these people with you when you begin your estimating process with homeowners. For this reason, you need to

Paper type	Adhesive type
Regular paper	Wheat or cellulose paste
Lightweight prepasted, paper-backed vinyl	Adhesive installed on back of paper
Lightweight nonpasted vinyl	Powdered vinyl adhesive
Cloth-backed vinyl	Premixed adhesive
Heavy cloth-backed vinyl	Premixed adhesive
Heavy paper-backed vinyl	Premixed adhesive

FIGURE 11.9 Types of wallpaper and adhesives.

- Don't apply wallpaper to unprimed drywall. It will be much harder to remove the wallpaper if the walls have not been covered with a primer paint.
- Don't paper over latex paint.
- Don't paper over a glossy surface.
- Don't paper over a wall that is peeling or in poor condition.

FIGURE 11.10 Wallpaper warnings.

- Do you have an adequate supply of wall covering from the same manufacturing lot number?
- Have you prepped the walls?
- Are all necessary tools available?
- Have electrical covers been removed?
- Have you read the instructions for the product you are installing?

FIGURE 11.11 Wallpapering checklist.

gather some background information to talk intelligently about wallpaper (Figures 11.10 and 11.11).

Vinyl-coated. Vinyl-coated wallpaper can be used in any room where excessive humidity is not a problem. The price for this type of wallpaper varies a great deal. Hanging vinyl wallpaper is not particularly complicated, and this is a popular type of wallpaper.

Wet-look Vinyl. Wet-look vinyl is often used in areas where moisture is present. Rooms like bathrooms, kitchens, and laundry rooms lend themselves to wet-look vinyl. The cost for this paper is moderate, but lining paper should be factored into the cost. If any wall being covered has imperfections on it, they will likely show through unless a lining paper is used.

Lining Paper. Lining paper is used to give a good, smooth surface for finish papers to be applied to. The inexpensive cost of lining paper does not make it a major financial inconvenience. However, if it is not used, the finished product, even of a high-quality paper, can turn out poorly. Wheat paste is normally used to apply lining paper.

Paper-backed Vinyl. Paper-backed vinyl wallpaper is a great all-purpose wallpaper. It is suitable for use in high-traffic areas and high-humidity areas. Prices on this paper range considerably. All in all, paper-backed vinyl is very hard to beat.

Cloth-backed Vinyl. Cloth-backed vinyl can be used in the same rooms as paper-backed vinyl, but there are more drawbacks to the cloth-backed paper. Price is one of them; this paper is expensive. Stiffness can also be a problem with cloth-backed vinyl. I'm not sure why anyone would opt for cloth-backed paper when paper-backed vinyl is readily available.

Foil. Foil wallpaper works well in rooms when constant cleaning is needed, such as bathrooms and kitchens. Lining paper should be used whenever a foil paper is to be installed. You should also be aware that the price for foil paper might be a bit steep for some budgets (Fig. 11-12).

Burlap. Burlap-style wallpaper can be beautiful, but it should not be used in rooms where it will take a beating. If a room will be damp, like in a bathroom, burlap should not be used. Kitchens are another place where a burlap-type paper shouldn't be used. The grease in a kitchen can wreak havoc with a grass-cloth or burlap-type paper.

Accent Paper. Accent paper has become very popular. How many kitchens have you seen with pineapples spreading around the tops of walls? Some kitchens have designs stenciled into them, and others use accent papers to achieve a similar look. Both methods are effective. There are dozens and

Roll type	Capacity (in sq. ft.)
American single roll	36
European single roll	28
Metric single roll	20

FIGURE 11.12 Wallpaper roll types and capacities.

dozens of accent pieces available. The best way to become familiar with all of these options is to spend some time looking through catalogs.

Tile

Tile can be an excellent wall covering. It can also be used very effectively as an accent. When tile is installed as a backsplash in a kitchen, it is easy to clean. It's also attractive, but it tends to be expensive. Tile has been used in bathrooms for years (Figure 11.13). It is normally installed about half way up the walls and around bathing units. Trends have moved away from tile, except in expensive homes, but people still like it. The reason it doesn't show up as often is cost. Tile and its installation is not cheap.

Ceramic tile comes in various shapes, a multitude of colors and a wide variety of designs. There is so much that can be done with it that it could boggle the mind. I incorporate tile into many of my jobs (Figure 11.14), and I suggest that you consider doing the same thing.

Tub Surrounds

Tub surrounds are normally available in plastic and fiberglass. I prefer the fiberglass ones. Many of my bathroom jobs have involved the replacement of

Tile	Sizes
Ceramic tile	Sizes range from 1-inch squares to 12-inch squares Most tiles have a thickness of $\frac{5}{16}$ inch
Ceramic mosaic tile	Sizes range from 1-inch squares to 2-inch squares Also available in rectangular shapes, generally 1"-×-2" Average thickness is $\frac{1}{4}$ inch
Quarry tile	Sizes range from 6-inch squares to 8-inch squares Rectangular tile is available in 4"-×-8" size Typical thickness is $\frac{1}{2}$ inch

FIGURE 11.13 Tile sizes.

Type of tile	Size of tile	Maximum joint width	Minimum joint width
Ceramic mosaic	$2\frac{3}{16}$" or less	$\frac{1}{8}$ inch	$\frac{1}{16}$ inch
Ceramic	$2\frac{3}{16}$" to $4\frac{1}{4}$"	$\frac{1}{4}$ inch	$\frac{1}{8}$ inch
Ceramic	6" × 6"	$\frac{3}{4}$ inch	$\frac{1}{4}$ inch
Quarry	All sizes	$\frac{3}{4}$ inch	$\frac{3}{8}$ inch

FIGURE 11.14 Recommended joint widths for wall tile.

faulty tub surrounds and leaking tile. As you probably know, the grouting between tile often allows water to seep past it. It is possible to regrout tile, but my experience shows that most people prefer to have the leaking tile replaced.

My standard procedure for the replacement of a bad tub surround is to strip the walls down to the bare studs. Then I install water-resistant drywall and install a quality, fiberglass tub surround. My favorite type is fairly thin and is applied directly to drywall with an adhesive. I've used this type of surround for probably 15 years without ever having a call-back on one.

You can buy cheap tub surrounds. I've seen them retail for less than $50. The type I use wholesales for between $250 and $300. This is a lot more money, but I feel the expense is worthwhile. Cheap surrounds tend to fall off the walls that they are applied to. I suppose this is due to the adhesive, but I'm not sure. What I am sure of is that the high-quality fiberglass units I install don't fall off.

Some fiberglass surrounds are meant to be installed directly to bare studs. I've used these types of units, and they work fine. I have no complaints against these enclosures. Still, I find the glue-on type to be better accepted among homeowners. I don't know why this is, but it's proved to be true for me.

If you really want to spend a lot of money on a tub surround, you can invest in some of the thick, marble-like surrounds available. These surrounds are very attractive and extremely durable. They are a little harder to work with, and they are a lot more expensive, but there are times when these surrounds are the ideal choice.

LAY IT OUT

When you are talking about wall coverings with your customers, lay it all out for them. Carry samples with you if you want, but one way or the other, be prepared to discuss the abundance of options available. Your performance in this area can have a direct affect on the success of your business. Learn what you need to know so that you can provide customers with enough information to make informed decisions.

CHAPTER 12
INTERIOR AND EXTERIOR DOORS

There are few major jobs done where doors of some type are not involved. You might be installing new doors where doors have never been before. Your job might require replacing a standard entry door with a sliding-glass door. A French door might be on your list of things to do. Doors are a part of most big jobs.

Compared to other aspects of building and remodeling, the installation of doors is not real high on the list of complex tasks to undertake. However, installing a door can involve a lot more than you anticipate. This is especially true if you are installing a replacement door of a size different than that of the existing unit.

There is, of course, more to doors than just the physical installation of them. As a professional contractor, you are also the property owner's consultant. When asked for recommendations on types of doors, you are expected to have some quality information to offer. This is perhaps one of the weak points of many contractors.

EXAMPLE SCENARIOS

As we progress in this chapter, we are going to discuss different types of doors. We will also talk about common problems that sometimes arise around door installations. However, to get us started, we are going to test your thought process. For the sake of this exercise, assume that I'm a homeowner and that you are a remodeling contractor who is talking with me about some work I want done. I will give you a few scenarios to respond to, and we will assess your performance at the end of each example.

Sliders Versus Gliders

This first story will cover the issue of sliders versus gliders. In this situation, I am an elderly person who wants to replace my old single entry door with a much larger, primarily glass door. As part of this project, you will be required to build an exterior sun deck for the new door to open onto. Our conversation has just turned to doors, so let's see what you will recommend.

I'm advancing in age, and I have limited arm strength. My goal is to have a new door installed that will let in a lot of light. I want a door that will provide adequate energy efficiency, but it must be easy for my arthritic hands to open. I don't do well with regular round door handles. After doing some personal research, I've come to the conclusion that I probably want a vinyl-clad, high-quality, sliding-glass door. One with a screen, of course. What are your feelings on this issue?

As a professional, you have a certain obligation, in my opinion, to help this customer with what can be a difficult decision. Consider the circumstances surrounding the customer for a moment. What recommendations would you make under these conditions? Would you take the easy route and go along with what the customer has expressed an interest in? Should you make the customer aware of additional options that might better suit the needs of a slightly restricted, elderly person? I believe this is a time to make the customer aware of gliding doors.

Gliding doors look like sliding-glass doors. The appearance of a glider will satisfy the look that the customer is after. Not only will the gliding door meet the requirements for style and appearance, it will be easier for the customer to open and close. The handle on a glider or slider will be sufficient for someone with stiff fingers to use, but you might have to install a second handle on the side of the door where only a finger groove is provided. This is a minor point to anyone who is not afflicted with some type of disability, but it can be a big advantage to someone who suffers from arthritis. Gliding doors open and close with much less effort than that required to move a sliding-glass door. This is an advantage for everyone, but especially for someone with limited arm strength.

Because energy efficiency is a concern of the customer, you might offer information on terrace doors. The look of these doors is not the same as a slider or glider, but a lot of light is available with this design. Installing a lever-type handle on the door will overcome the problem this customer has in dealing with round knobs, and efficiency should be better with a terrace door.

Can you see from this example how there is more to the installation of doors than just carpentry work? Any good carpenter can install a door, but it takes a thoughtful and experienced contractor to make contributions to customers that are above and beyond the normal level.

Six-panel Doors

In this example, I am a customer who has my heart set on installing all new six-panel, interior doors. This work will be done in conjunction with other interior remodeling work, which will require the replacement of existing interior trim. I'm an average homeowner, so money is a factor, but it is not the only motivator in my decision-making process. What type of door would you recommend?

As this remodeler, you have two types of doors to offer me. You can sell me pine doors or molded doors. Which one are you going to work with me on? Ah, you need more information, don't you? Your first question to the homeowner in this example should be in regard to whether the doors will be painted or stained. If the doors are to be stained, pine doors are the logical choice. When painting is planned, either door will be suitable for the job, and molded doors will be less expensive, while providing the same basic appearance.

Sometimes you have to ask questions before you can give customers good answers. Let's assume that the homeowner in this story was willing to accept molded doors, but you just assumed that pine doors were what the customer would want. You would bid the job with pine doors. Now, what would happen if a second contractor came along behind you and talked more extensively with the homeowner. This contractor might discover that molded doors are fine with the property owner. When the second contractor puts in a bid for the same basic work that you gave a price for, your price is going to be high, assuming that the price of the doors in the estimate are the only variable. If you don't take the time to discuss jobs thoroughly with your customers, you might be doing them a disservice, as well as yourself.

Closet Doors

Our next story has to do with closet doors. You have a customer who is remodeling extensively, and the work will require the replacement of all existing closet doors. There will also be some new closets built that will need doors. Your customer has always been used to sliding closet doors and doesn't consider any other possibilities. You are given a set of plans and specifications for the job, and it clearly states the installation of sliding doors for all closets. What are you going to do?

Most contractors, in a situation like the one just described, will bid a job as per plans and specs. You can do this without any guilt, but you might be able to improve your odds of a sale if you voice your opinion. What's wrong with installing sliding doors on a closet? You should know the answer to this question without even thinking about it. When sliding doors are used, only one-half of the closet is accessible at any one time. A set of bifold doors will allow full access to the closet, so why not use them? This is the question I would pose to the customer.

When you step out of the crowd and show customers your knowledge and your concern for their well being, you are well on your way to becoming a more successful remodeler. Something as simple as pointing out the good points of bifold doors can be all it takes to knock your competition out of the running. Remember this, you are a remodeling contractor, not a professional salesperson, but without sales, there is no remodeling work to do.

Traffic Pattern

This story has to do with a homeowner who is worried about the traffic pattern around the kitchen, living room, and dining room. You are building a new addition on the home that will serve as a dining room. Because the house has never had a formal dining room, all meals were prepared in the kitchen and served at an eat-in location. The new dining room will be located in such a way that a brief journey through the living room will be required to get to it.

Both homeowners are concerned about the new door being installed between the kitchen and the living room. They don't want a cased opening, because they don't want guests to have a clear view of the kitchen when they are seated in the living room. However, they are concerned that a door will cause problems, due to its swing. If the door swings in, it will block the appliances in the kitchen. A door that swings out, into the living room, will be cumbersome. What can you recommend?

A simple, cost-effective solution to this problem would be a set of swinging doors, the types depicted in old saloons. These doors, strategically placed, would block direct vision of the kitchen counter and work area. While dirty dishes and such would be obscured from view, the swinging doors would allow a flowing, open look to the home's design. This could certainly be the ideal solution, but suppose the customer wanted something a bit more formal? What would you suggest at this point? How about a six-panel pocket door?

The installation of a pocket door will not create a problem like a swinging door would. The pocket door will store in the recess of a wall when opened, and the six-panel door will look elegant when closed. A pocket door will also serve to block all view of the kitchen. If a swinging door is not right, a pocket door should do the trick.

TYPES OF DOORS

By now, you should be getting an idea of how the use of different door styles and accommodate the various needs and desires of your customers. What works for one customer might not suit another. Once you have a full arsenal of door products to offer, you can fill every void and niche. With this in mind, let's move onto the different types of doors available.

Exterior Doors

Exterior doors come in all shapes and sizes (Figure 12.1). They are also available in a wide variety of materials. Matching the right door to your customer's need is the first step in a successful installation. We have looked at some examples of how this type of matching might be done, but we haven't explored the many options available for you to offer your customers. Let's do that now.

Height	Width
80"	36"
84"	36"
80"	34"
84"	34"
80"	32"
84"	32"

FIGURE 12.1 Stock sizes of exterior doors.

Standard Entry Doors. We often hear people talk about standard entry doors, but what is a standard entry door? There are so many types of doors available for use as an entry door, it hardly seems possible to correctly term any one door as a standard entry door. In general, a standard entry door, in my opinion, is a solid door that has a width of 3 feet. Even after making this stipulation, there are a multitude of doors that fit the mold of a standard entry door. To clear this up, let's dig a little deeper.

Wood Doors. Wood doors have long been installed in homes. There was a time when they were an industry standard for residential construction. They are, however, losing some ground in today's construction market. While not as popular as they once were, wood doors still command their share of attention. This attention can come at a substantial price.

Wood doors are often attractive, and they can be quite ornate. The doors are able to be painted or stained, and this advantage alone sometimes sells them. With the variety of styles available for wood doors, it is easy for one's mind to become boggled. All you have to do is skim through a supplier's catalog to see the pages after pages of doors offered in wood construction. Certainly, this wide selection is another selling feature for wood doors. However, as good as wood doors can be, they harbor their downsides.

It is not uncommon for a wood door to swell when it becomes damp. I've seen wood doors swell to a point where they could not be opened, even with somewhat excessive force. This, of course, is not the type of situation most people want to encounter when they are late for work or their house is on fire. So, we have to say that swelling is a strike against wood doors.

How do wood doors stack up in the security department? They are unquestionably more secure than a glass door, but not as dependable as a steel door. This might or might not be a factor when offering a type of door to your customer, but it is an issue you should at least keep in mind.

Will a wood door rot? Of course it will if it is not protected properly. When you compare the longevity of a wood door to a steel door, you are going to have a difficult time showing any advantage for the wood unit. This factor might influence a customer's buying decision.

Can a wood door be stained? Yes, and this is one of its stronger selling points. Steel doors are good in many respects, but they are worthless to the homeowner who wants to have a stained door. Wood doors don't own the do-main of stained doors, but they certainly take a lion's share of this market.

Where does a wood door stack up in a review of energy efficiency? Again, wood is a better insulator than glass, but a wood door cannot compete with an insulated steel door. From an insulating point of view, wood doors are okay, but they are nothing to rave about.

Price is often a consideration in the purchase of a door. Wood doors range widely in price, but they tend to be affordable. By affordable, I mean that a wood door is not so expensive that it can be installed only in upper-crust homes, but neither is it the most competitive type of door when it comes to a price war.

When should a wood door be used as an entry door? Anytime a customer is willing to pay for one. Now really, you should take a more serious approach to the question of when to use a wood door. One founded reason would be if the door is to be stained. However, a fiberglass door might be a better alter-native. I think appearance is the major factor to focus on when evaluating a wood door. If your customer is seeking a certain look, wood might be the best way to satisfy the customer's desire. In my opinion, appearance and staining are the two most logical reasons to opt for a wood door.

Fiberglass Doors.   Fiberglass doors are relatively new on the market. These innovative doors, however, bring much to the bargaining table. Some of them can be stained to achieve similar results as you would with a wood door. A fiberglass door will not swell, and it should have a better insulating value than a wood door. Security is good with a fiberglass door, and about the only real drawback I can think of is the price you might pay for such a door. I must admit, I've never installed a fiberglass door. My experience has in-cluded shopping these doors for customers who wanted a stainable alternative to wood, but the cost scared my customers away. From everything that I've learned about them, fiberglass doors do seem to offer many features worth considering.

Steel Doors.   Insulated steel doors are the mainstay of my business when it comes to main entry doors. Six-panel, embossed steel doors are both very affordable and quite popular. Their insulating qualities are good, and they don't swell up like a wood door can. Just the fact that they are made of steel imparts a sense of security, and customers seem drawn to them. Their low cost doesn't hurt their appeal, either. As long as a customer is willing to live with a painted door, I think a steel, insulated door is hard to beat.

Glass.   A main entry door that contains glass can be a security risk. It is certainly a place for some heat loss, but this is not as big of a deal as some people make it out to be. How many houses have you seen that didn't have windows? If the glass in windows is acceptable from a heat-loss perspective, why shouldn't the glass in doors fall under the same guidelines? The security issue is somewhat different. A door with a lot of glass in it, or even just a lit-tle glass near the locking arrangements, is a liability when compared to a solid door. There is no good defense around this issue. However, what is to stop a criminal from breaking out a window if access is wanted through a glass obstacle? Let's face it, if a burglar wants into a house badly enough, there are very few security measures that are feasible that will prevent entry.

Some people feel more secure when an entry door has some glass in it. The glass allows them to see who is on the other side of the door. Sidelights are often used for this purpose, but they to can pose some security risks. If a key-operated dead-bolt lock is installed on a door, a small pane of glass broken out is not enough to allow entry to an intruder. Regardless of your personal feelings, or mine, there are customers who demand entry doors that contain glass. When this is the case, there is a lot to offer a customer.

I don't want to lump glass doors all together here. There are so many types of doors that consist mainly of glass that I would prefer to discuss them individually. For the sake of this section, let's limit glass doors to be those that include look-out glass. An example could be a door with a fan-shaped glass top.

Aside from seeing who is on the other side of a door, there is another valid reason for installing an entry door that contains glass. The reason is natural light. Doors that have glass in them brighten up a foyer or home. The same is true of transoms installed over doors. Architecturally, glass-filled doors and glass transoms can do a lot to set a house apart from others in the neighborhood. There are certainly some good reasons for installing doors that contain glass.

Nine-lite Doors. Nine-lite doors, doors with nine glass panes in the upper half of them, are very popular for side and back entrances. This particular type of door is well received when it joins a kitchen to the exterior of a home. The construction material of a nine-lite door can vary, but it is the bright natural light offered from this type of door that attracts customers. Due to the design of this type of entry door, security is sacrificed for appearance and the functional use of natural light. Still, these doors rank high in customer popularity.

Sliding-glass Doors. Sliding-glass doors (Figure 12.2) are frequently used as entry doors to specialized portions of homes. If a house is equipped with a

Glass size (in inches)	Frame size (width × height, feet and inches)	Rough opening (width × height, feet and inches)
33 × 76¾	6-0 × 6-10¾	6-0½ × 6-11¼
45 × 76¾	8-0 × 6-10¾	8-0½ × 6-11¼
57 × 76¾	10-0 × 6-10¾	10-0½ × 6-11¼
33 × 76¾	9-0 × 6-10¾	8-0½ × 6-11¼
45 × 76¾	12-0 × 6-10¾	12-0½ × 6-11¼
57 × 76¾	15-0 × 6-10¾	15-0½ × 6-11¼
33 × 76¾	11-11 × 6-10¾	11-11½ × 6-11¼
45 × 76¾	15-11 × 6-10¾	15-11½ × 6-11¼
57 × 76¾	19-11 × 6-10¾	19-11½ × 6-11¼

FIGURE 12.2 Measurements for sliding-glass doors.

deck, there is a good chance the door leading from the home to the outside play space will be a sliding-glass door. These doors are used in sun rooms, off of breakfast areas, and sometimes in master bedrooms. Sliding-glass doors can't be beat if the goal is to brighten up a room, but there are trade-offs.

Many sliding-glass doors rate poorly in energy efficiency. Much of this has to do with the volume of glass involved in the door's construction. However, there are many inexpensive doors that don't fit up well, therefore, allowing heat to escape and cold air to infiltrate. Cheap sliders often condensate terribly. This can lead to moisture problems in carpeting, and it is not unusual for these doors to frost up in winter.

The mere act of opening or closing a cheap slider can be more than some people can handle. Security is another issue. It is impossible to create a good sense of security with a sliding-glass door. Special devices that wedge the door shut protect against some unauthorized entries; however, because the door is made almost entirely of glass, there is no protection from a criminal with a hammer.

Even with the many troubles surrounding them, sliding-glass doors remain relatively popular. If you are going to sell a customer a cheap slider, make sure the customer is aware of what the trade-offs for the low-priced model are. My experience has shown that stock sliders can range from less than $300 to well over $1200. This is quite a price span, and there are some good reasons for it. You should take the time to explain the features and benefits of various doors to your customers. Point out why some doors cost more than others, and let the customer decide what features are worth paying for.

Gliding Doors.   Gliding doors, as we have already seen in one of the earlier examples, mirror sliding-glass doors in appearance and general function. The biggest advantage to a glider is the ease with which it can be opened and closed. Depending upon the physical condition of your customer, this can be a big benefit in itself.

Terrace Doors.   If sliding-glass doors have been threatened in the marketplace, it has been largely due to terrace doors. These hinged doors provide a quantity of light similar to that of a slider, but they do so in a more energy-efficient package. There are also easier to operate. Terrace doors have one fixed panel and one panel that swings open. They can be filled with glass, but unlike a slider, terrace doors will accept slip-in mullions that give the appearance of individual panes of glass. This is often considered to be an upgrade over a solid glass panel.

Because terrace doors operate much like any other entry door, they can be fitted with dead-bolt locks, and their weatherstripping is often more effective than that of a sliding door. Terrace doors are not perfect for all remodeling jobs, but they do seem to command a lot of favorable attention.

French Doors.   French doors are not used too often as entry doors, but they can be. These doors tend to be very expensive. Their primary appeal is that both panels of the double door open, unlike those of a terrace door, where only one side opens. French doors usually have individual panes of glass that run from the top of the door to the bottom. Security is obviously difficult with a French door, and so is protection against heat loss. These doors see more use as interior dividers than they do as exterior doors.

Interior Doors

When you are ready to discuss interior doors with your customers, the list of options will not be as long as the one associated with exterior doors. This, however, is not to say that there is not a lot to consider. Construction material will not be as much of an issue, and neither will energy efficiency. Security risks will diminish, and so will some of the glass options. However, this still leaves wood doors, molded doors, bifold doors, pocket doors, sliding doors, and so forth. The doors you guide your customers to for the interiors of their homes can affect the appraised value of their remodeling efforts. If for no other reason, this is reason enough to take door selection seriously (Figure 12.3).

Molded or Wood. One of the first questions you might be faced with is a customer who can't decide whether to go with a molded or wood door. There is one quick test that might put this discussion to a quick end. Molded doors will not take stain effectively. If your customer wants stained doors, molded doors should be dropped from consideration. When the doors will be painted, your debate must continue.

Price might very well be the second criteria to use when trying to decide between molded and wood doors. Pine doors can cost twice what a molded door costs. I'm speaking of six-panel doors. Molded doors, when painted, give a good showing for themselves. In my opinion, a molded door that has been painted is every bit as attractive as a wood door that has been painted. I've used molded doors on my jobs for years, and I've used them in the various homes I've built for myself. Through all of this use, I've never found anything to complain about with them.

There is a segment of the buying population that feels they only get their money's worth with wood doors. I disagree with this, but I admit that there are times when it is senseless to argue the issue. If the old saying about customers

Height	Width
80"	24"
80"	28"
84"	28"
80"	30"
84"	30"
80"	32"
84"	32"

FIGURE 12.3 Stock sizes of interior doors.

always being right is true, you are sure to paint some wood doors in your remodeling career. To me, this seems like a waste of money, but it does pay to give your customers what they want.

Luan Doors. Up to now I've been discussing six-panel doors, but what about luan doors? These flat, hollow-core doors will accept stain, and they are inexpensive. I've installed them in numerous starter homes and in remodeling jobs were budgets were tight. They are a very plain type of door, but there is no question that they are serviceable, and they do allow the option of staining. From a personal point of view, I prefer six-panel doors, but flat luan doors do have their place.

Sliding Doors. Interior sliding doors are normally used in conjunction with closets. These are typically flat-surfaced, hollow-core doors. They are inexpensive and effective. The biggest complaint I'm familiar with surrounding these doors is that only one-half of a closet can be accessed at any one time when they are installed. This is a reasonable grievance. I can't remember ever installing any of these doors.

Bifold Doors. Bifold doors are generally considered to be head and shoulders above sliders when used for pantries, laundry nooks, and closets. The big advantage to bifold doors is that when they are open, the space behind them is nearly 100% accessible. These doors are so well accepted that I rarely bother with any other type that might be a competitor.

Pocket Doors. Pocket doors are special-use doors. I have installed them for bathrooms, kitchens, studies, and probably some other uses that are not coming to mind. These doors store in a wall when not in use, and because they don't swing open, they never interfere with the use of a room. Pocket doors are not practical for all uses, but they certainly do a nice job under special circumstances. I should mention that these doors are available in a six-panel configuration.

THE ACTUAL WORK

The actual work involved with replacing or installing a door is not much of a problem for experienced carpenters. There are, however, a few little problems that can pop up. It is also possible to save some time on the job by purchasing the right types of doors. We will finish out this chapter by talking about potential problems and time-saving techniques.

Pre-hung Doors

Pre-hung doors are, in my opinion, the only way to go. Why would you ever buy a door and build the complete door unit on site? Sure, there might be a rare occasion when a custom-made door unit is needed, but 9 times out of 10, a pre-hung door will do just fine. Think of all the time that is saved with a pre-hung door. They cost a little more than a slab door and the components needed to make a complete unit, but the time saved in labor more than offsets the cost. It is beyond me why any contractor would prefer to work harder instead of smarter, but there are people in the trades who would. I recently encountered two of them, and it cost me some money.

In addition to being a remodeler, I'm also a home builder. I recently built a new house for myself. The carpentry crew I used was one that I had used on two other houses, but we were not as accustomed to working together as some of my crews and I have been. When it came time to order my interior doors and trim, a big mistake was made. Some of the fault, okay—all of the fault, rests on my shoulders as a lack of good communication. As the general contractor, whatever goes wrong on a job is your fault, and this means that I must assume full responsibility for the stupidity revolving around the ordering of my interior doors.

I ordered pre-hung, molded, six-panel doors that were pretrimmed with split jambs. This is one of my standard types of interior doors. Unfortunately, I was using my new house as a testing ground for this particular carpentry crew. I wanted to see if they had the ability to be a stand-alone, independent representative for my company. They didn't pass the test, in more ways than one.

Anyway, I asked the crew leader to give me a take-off for the interior doors and trim. Being pressed for time in many areas, I failed to do my own take-off. This is where my stupidly rears its ugly head. My crew leader provided an itemized take-off, as requested, and I ordered the material. It never occurred to me that anyone would order pre-hung doors that were not fitted with pretrimmed split jambs. My guy did.

When the casing was delivered to my job, I went to work painting it furiously. The casing just happened to be the first part of the trim that Kimberley, my wife, and I painted. By the end of the week-end, we had most of the trim painted. A day or so later, the doors arrived. My crew leader took one look at the doors and got an expression on his face that told me something was wrong. Guess what happened?

My carpenter, who has built a good number of houses, had ordered enough casing stock to trim the doors on site. He never considered that I was going to order pretrimmed doors. I never thought about anyone planning to install this type of door in any other way. The result was my having special-ordered doors that were pretrimmed and a garage full of painted casing, much of which wouldn't be needed. Kimberley and I were more than a little upset. If you've priced Colonial casing lately, you will understand why we were not happy.

I've told you this story not only to show you that I'm human, but to illustrate that not all builders and carpenters think alike. I can't imagine wanting to spend the time needed to trim each door individually when doors can be ordered with pre-fab trim packages. Obviously, not all carpenters share my opinions. I stick by my guns though, there is no more cost-effective way to order doors than to have them shipped pre-hung and pre-trimmed.

New Doors

Installing new doors in rough openings that you have created is not much of a job. Any experienced carpenter can handle this work. However, remodelers are often faced with circumstances that are less than perfect. Someone who is installing new doors on a new-construction job or a new addition shouldn't run into any major trouble. Take this same task and associate it with remodeling and cutting new doors into old walls, and you have potential for problems.

It is common to use shims when installing doors under the best of conditions. The shims are usually thin and provide minor modifications of a rough opening to insure a good door installation. However, what happens when a standard shim isn't enough? Many old homes have shifted and settled to a point where finding a plumb wall is nearly impossible. I've seen jobs where the existing framing was so far off that the carpenters had to use some substantial lumber as shims, of a sort. When this has happened, the carpenters fumed and fussed about the walls. They worked for quite a while to bring existing studs into an acceptable condition. Personally, I never understood why they went to so much trouble working with old framing. Is there an alternative? Yes.

If you open up a wall to install a new door, you are going to have to do some framing. The wall is going to need repair before the job is done, and time is money. What would you do under these conditions? Well, as I said, I've seen carpenters take detailed measurements and trim lumber meticulously to make a plumb opening. This, in my opinion, is a waste of time. Why not just install new studs to create the opening, even if it means removing one or two old studs? This approach will generally prove to be faster and easier than trying to salvage the use of old studs.

Enlarged Openings

Remodelers frequently have to replace existing doors with larger doors or different types of doors, this normally means making enlarged openings. The work here is simple enough, until some obstacle is encountered. You might find that a plumber ran a vent or drain pipe up the wall in an area where you now have to install a door. There is a good chance that electrical wires will be in your way. Either of these situations will require the services of a licensed trade, and your cost on the job will go up. If you will be enlarging existing door openings, budget some money for relocation work associated with licensed trades.

STAYING OUT OF TROUBLE

Staying out of trouble is not difficult if you look ahead and plan your work. Hanging new doors will not be troublesome if your rough framing has been done properly. Assuming that the opening is the right size and that it is plumb, putting in a new door will not take long.

Most contractors lose time installing doors because they, or their predecessor, did not frame the rough openings properly. It is a very good idea to check all openings before drywall is hung. If you catch a framing problem before the walls are covered, changes will not be especially difficult to make. If you wait until the job is nearly done and you are in the process of hanging new doors to discover an error in framing, your finish work can take a lot longer than you planned for.

CHAPTER 13

ROOFING SYSTEMS AND COVERINGS

Roofing systems and coverings are included in just about every type of building project. Whether you are a remodeler or a builder, you will probably spend a good deal of time working with roofing. What can go wrong with a roof structure? Plenty. I've seen rafters turned to sawdust by wood-infesting insects. Granted, this should not be your problem; however, if you fail to do a thorough investigation before quoting a price and offering a proposal, you could wind up paying to replace a lot of roof material.

Bugs aren't the only enemy in an attic. Houses without proper attic ventilation can rot their roof structures quickly. When the attic of a home is not insulated and vented properly, condensation can form at an alarming rate. The roof sheathing will become saturated, and the rafters or trusses will begin to take on water. This type of problem usually exposes itself in the form of leaks coming through ceilings. I have seen such cases where water was literally dripping off of the roof sheathing and soaking the attic insulation. It doesn't take long for this type of problem to do some extreme damage. If you happen to commit to doing a job involving such a situation, without prior knowledge of the defects, you can be in big trouble.

Most roof systems don't just deteriorate for no reason. Before major structural damage is done, there has to be some type of defect. It can be condensation or bugs, but it could be a regular old roof leak. Of course, this type of problem isn't going to destroy an entire roof system, like bugs and condensation can. To protect yourself, you have to do a full inspection of the roof system before committing to any work involving it.

One effective way to test a roof structure is done with the use of a probing tool. This can be a knife, a screwdriver, an awl, or any other similar item. By poking the wood with such a tool, you can tell if the lumber is solid. I have seen structural timbers that looked fine on the outside but that were gutted on the inside. This was the work of wood-infesting insects. Don't count on just what you can see; probe randomly to see that the entire system is satisfactory.

TRUSSES

Trusses (Figures 13.1 through 13.26) make a good roof system, and they are economical to use. When building new construction, trusses are great, but they

Chord size: 2-×-4 top and 2-×-4 bottom	
Pitch	Span (feet)
2/12	22
3/12	29
4/12	33
5/12	35
6/12	37

FIGURE 13.1 Typical truss spans (55 PSF with 15% duration factor).

Chord size: 2-×-6 top and 2-×-4 bottom	
Pitch	Span (feet)
2/12	28
3/12	39
4/12	46
5/12	53
6/12	57

FIGURE 13.2 Typical truss spans (47 PSF with 33% duration factor).

Chord size: 2-×-6 top and 2-×-6 bottom	
Pitch	Span (feet)
2/12	32
3/12	51
4/12	56
5/12	60
6/12	62

FIGURE 13.3 Monopitch truss spans (55 PSF with 33% duration factor).

Chord size: 2-×-4 top and 2-×-4 bottom	
Pitch	**Span (feet)**
2/12	25
3/12	33
4/12	37
5/12	40
6/12	41

FIGURE 13.4 Monopitch truss spans (55 PSF with 33% duration factor).

Chord size: 2-×-6 top and 2-×-4 bottom	
Pitch	**Span (feet)**
2/12	23
3/12	31
4/12	39
5/12	45
6/12	51

FIGURE 13.5 Monopitch truss spans (55 PSF with 15% duration factor).

Chord size: 2-×-4 top and 2-×-4 bottom	
Pitch	**Span (feet)**
2/12	22
3/12	30
4/12	33
5/12	35
6/12	37

FIGURE 13.6 Monopitch truss spans (55 PSF with 15% duration factor).

can give a remodeler a lot of problems. Let's say that you are giving an estimate for installing a dormer on a Cape-Cod style home. The upstairs of the house is finished, and there is no access into the knee-walls. Your customer wants a shed dormer installed so that a bathroom can be added in the upstairs hall.

Chord size: 2-×-6 top and 2-×-6 bottom	
Pitch	**Span (feet)**
2/12	34
3/12	45
4/12	50
5/12	53
6/12	56

FIGURE 13.7 Monopitch truss spans (55 PSF with 15% duration factor).

Chord size: 2-×-6 top and 2-×-4 bottom	
Pitch	**Span (feet)**
2/12	25
3/12	35
4/12	42
5/12	48
6/12	53

FIGURE 13.8 Monopitch truss spans (55 PSF with 33% duration factor).

Chord size: 2-×-6 top and 2-×-6 bottom	
Pitch	**Span (feet)**
2/12	44
3/12	57
4/12	63
5/12	67
6/12	68

FIGURE 13.9 Monopitch truss spans (47 PSF with 33% duration factor).

Chord size: 2-×-4 top and 2-×-4 bottom	
Pitch	Span (feet)
2/12	28
3/12	38
4/12	42
5/12	44
6/12	45

FIGURE 13.10 Monopitch truss spans (47 PSF with 33% duration factor).

Chord size: 2-×-6 top and 2-×-4 bottom	
Pitch	Span (feet)
2/12	28
3/12	38
4/12	46
5/12	52
6/12	57

FIGURE 13.11 Typical truss spans (47 PSF with 33% duration factor).

Chord size: 2-×-4 top and 2-×-4 bottom	
Pitch	Span (feet)
2/12	28
3/12	37
4/12	41
5/12	44
6/12	44

FIGURE 13.12 Typical truss spans (47 PSF with 33% duration factor).

Chord size: 2-×-6 top and 2-×-4 bottom	
Pitch	Span (feet)
2/12	25
3/12	34
4/12	42
5/12	48
6/12	53

FIGURE 13.13 Typical truss spans (55 PSF with 33% duration factor).

Chord size: 2-×-4 top and 2-×-4 bottom	
Pitch	Span (feet)
2/12	25
3/12	33
4/12	37
5/12	39
6/12	41

FIGURE 13.14 Typical truss spans (55 PSF with 33% duration factor).

Chord size: 2-×-6 top and 2-×-6 bottom	
Pitch	Span (feet)
2/12	39
3/12	50
4/12	55
5/12	59
6/12	62

FIGURE 13.15 Typical truss spans (55 PSF with 33% duration factor).

Chord size: 2-×-6 top and 2-×-4 bottom	
Pitch	Span (feet)
2/12	23
3/12	31
4/12	39
5/12	45
6/12	51

FIGURE 13.16 Typical truss spans (55 PSF with 15% duration factor).

Chord size: 2-×-6 top and 2-×-6 bottom	
Pitch	Span (feet)
2/12	34
3/12	44
4/12	49
5/12	53
6/12	55

FIGURE 13.17 Typical truss spans (55 PSF with 15% duration factor).

Chord size: 2-×-4 top and 2-×-4 bottom		
Top chord pitch	Bottom chord pitch	Span (feet)
6/12	2/12	36
6/12	3/12	31
6/12	4/12	24

FIGURE 13.18 Scissor truss spans (55 PSF with 33% duration factor).

Chord size: 2-×-6 top and 2-×-6 bottom		
Top chord pitch	Bottom chord pitch	Span (feet)
6/12	2/12	54
6/12	3/12	48
6/12	4/12	36

FIGURE 13.19 Scissor truss spans (55 PSF with 33% duration factor).

Chord size: 2-×-6 top and 2-×-4 bottom		
Top chord pitch	Bottom chord pitch	Span (feet)
6/12	2/12	38
6/12	3/12	30
6/12	4/12	22

FIGURE 13.20 Scissor truss spans (55 PSF with 15% duration factor).

Chord size: 2-×-4 top and 2-×-4 bottom		
Top chord pitch	Bottom chord pitch	Span (feet)
6/12	2/12	32
6/12	3/12	28
6/12	4/12	21

FIGURE 13.21 Scissor truss spans (55 PSF with 15% duration factor).

Chord size: 2-×-6 top and 2-×-6 bottom		
Top chord pitch	Bottom chord pitch	Span (feet)
6/12	2/12	48
6/12	3/12	42
6/12	4/12	32

FIGURE 13.22 Scissor truss spans (55 PSF with 15% duration factor).

Chord size: 2-×-6 top and 2-×-4 bottom		
Top chord pitch	**Bottom chord pitch**	**Span (feet)**
6/12	2/12	42
6/12	3/12	34
6/12	4/12	24

FIGURE 13.23 Scissor truss spans (55 PSF with 33% duration factor).

Chord size: 2-×-6 top and 2-×-6 bottom		
Top chord pitch	**Bottom chord pitch**	**Span (feet)**
6/12	2/12	61
6/12	3/12	54
6/12	4/12	42

FIGURE 13.24 Scissor truss spans (47 PSF with 33% duration factor).

Chord size: 2-×-6 top and 2-×-4 bottom		
Top chord pitch	**Bottom chord pitch**	**Span (feet)**
6/12	2/12	46
6/12	3/12	38
6/12	4/12	27

FIGURE 13.25 Scissor truss spans (47 PSF with 33% duration factor).

Chord size: 2-×-4 top and 2-×-4 bottom		
Top chord pitch	**Bottom chord pitch**	**Span (feet)**
6/12	2/12	40
6/12	3/12	35
6/12	4/12	27

FIGURE 13.26 Scissor truss spans (47 PSF with 33% duration factor).

Because the house is of a Cape design, you assume it is framed with rafters. This is a logical assumption, but it could be an expensive mistake for you. In this particular case, the upper level of the house was done with room trusses. This is fine in the home's present use, but adding a dormer can get tricky, because of the trusses.

I'm not saying it will be impossible to fulfill your contract obligation to build a shed dormer on the house, but I don't think you will make as much money as you had planned to. Because the upper level has been done with trusses, you should seek expert engineering advice on how to cut the trusses to allow for the dormer. This advice can get expensive, and because you didn't disclaimer trusses, and in fact never even asked about them, the cost of engineering fees is coming out of your profits.

Trusses are great to work with when you are doing new construction, but they can be a serious threat to your bank account when remodeling around them. You should always attempt to do a visual inspection of any aspects you will be required to work with. In a case like the example I've given you, where the trusses are not visible, you should protect yourself with a disclaimer clause in your contract. For example, if you had stated in your contract that the agreement was based on the house having a stick-built roof and that additional fees would be charged if the roof was found to be built with trusses, you'd be off the hook.

RAISING A ROOF

Raising a roof is a big job. This is certainly the type of work where you should protect yourself with written recommendations from design experts. The occasions when most remodelers raise a roof are rare, but they do come along. Some customer decides to turn a ranch-style home into a Cape-style home. A customer wants to do a major attic conversion, but insists on more headroom. Either of these situations can call for removing an existing roof and building a new one. Before you agree to get involved in this type of work, make darn sure you know what you are getting into and that you are prepared to go the distance.

Once you take a roof off of a person's house, you had better be able to get a new one back on quickly. Having a week of rain set in on a house without a roof can ruin your rating with insurance companies. Seriously though, there is a lot of preplanning needed for a roof removal. Weather is a prime concern, but it is not the only one, and it might not be the biggest one.

When it comes to raising a roof, you must first get rid of the old roof. This, in itself, is a substantial job. You must make arrangements to set up and maintain a safe perimeter within which to work. You don't want to be tearing off the roof and have a huge chunk of it fall on a homeowner, child, or delivery person. Neither do you want a heavy section of roof dropping on your customer's car.

Rafter size (inches)	Spacing (inches)	Pine (feet/inches)
2 × 4	12	11-6
2 × 4	16	10-6
2 × 4	24	8-10
2 × 6	12	16-10
2 × 6	16	15-8
2 × 6	24	13-4
2 × 8	12	21-2
2 × 8	16	18-10
2 × 8	24	17-104
2 × 10	12	24-0
2 × 10	16	23-8
2 × 10	24	21-4

FIGURE 13.27 Typical maximum spans for roof rafters.

Then, of course, you have to make arrangements for removing the debris from your job site. This might be done with the use of container services or a large truck. However one way or another, you have to get rid of the old roof.

There is also the consideration for damage that might occur inside the home. When your crews are banging and cutting, you can bet that some of the living space will be disturbed. Expensive vases could be knocked off shelves. Light fixtures could lose their globes. Ceilings are going to be damaged, and other situations are bound to develop.

Looking past the removal of the old roof, what are you going to do once it's gone (Figure 13.27)? How are you going to keep the house protected from wind, rain, and other falling precipitation? Are you going to set up some rough framing and cover it with a tarp? Should you remove the old roof in sections and replace it as you go along? All of these questions are pertinent, and as a remodeling contractor, you had better have answers for them. A slip up at any stage of this type of work could cost you more money than you have.

ROOFING MATERIALS

Roofing materials are pretty standard, in terms of what is commonly used to cover the roof structure of a home (Figures 13.28 and 13.29). Asphalt shingles are certainly the most commonly used roofing material on residential properties. Fiberglass shingles are, however, making a move to out-sell asphalt, and they might be reaching their goal in some areas.

Roofing type	Minimum slope	Life years	Relative cost	Weight (pounds/100 sq ft)
Asphalt shingle	4	15–20	Low	200–300
Slate	5	100	High	750–4000
Wood shake	3	50	High	300
Wood shingle	3	25	Medium	150

FIGURE 13.28 Roofing materials.

Material	Expected life span
Asphalt shingles	15 to 30 years
Fiberglass shingles	20 to 30 years
Wood shingles	20 years
Wood shakes	50 years
Slate	Indefinite
Clay tiles	Indefinite
Copper	In excess of 35 years
Aluminum	35 years
Built-up roofing	5 to 20 years

Note: All estimated life spans depend on installation procedure, maintenance, and climatic conditions.

FIGURE 13.29 Potential life spans for various types of roofing materials.

Type of roofing	Weight in pounds
Clay shingle tile	1000–2000
Clay Spanish tile	800–1500
Slate	600–1600
Asphalt shingles	130–325
Wood shingles	200–300

FIGURE 13.30 Approximate weights of roofing materials (based on 100 square feet of material installed).

Other types of roofing, such as cedar shakes (Figure 13.30), tiles, and slate are not installed on many new houses in today's construction and remodeling market. If you have never worked with a slate roof, you might be surprised at how different this roofing material is from other types that you might have dealt with. The same can be true of working with a tile roof. A shake roof is different from an asphalt roof, but shakes are easier to work with than tile or slate is.

Asphalt Shingles

Asphalt shingles have long been an industry standard in residential roofing. They have been installed for years, and they are still the industry leader in residential roofing materials. I've worked with asphalt shingles throughout my 20-year career, and I've never been disappointed by them. In my opinion, these shingles make an ideal roof, and they are easy to work with.

Many of your customers will probably ask you what type of roofing material you will recommend. Each job can have its own set of circumstances, but I suspect that you will either suggest asphalt or fiberglass shingles. When you make a recommendation, you should back up your opinion with facts. Have some manufacturer brochures available to give to your customers. Take some time to go over the brochures with your customers. Your opinion is very valuable to a homeowner, but I recommend that you let the customers make their own decisions.

If you push a particular type of roofing on a customer, the sale might backfire on you. For example, if I was a big proponent of fiberglass shingles and sold them on all of my jobs, I might be confronted with some unhappy customers down the road. If you go out on a limb by making specific recommendations, you might find that you will be held liable, in some form, before you are done dealing with your customers.

Fiberglass

Climatic conditions can have an affect on roofing materials (Figures 13.31 and 13.32), such as fiberglass shingles. For example, fiberglass shingles might

Material	Slope
Asphalt or fiberglass shingle	4 in 12 slope
Roll roofing with exposed nails	3 in 12 slope
Roll roofing with concealed nails 3" head lap	2 in 12 slope
Double coverage half lap	1 in 12 slope
Lower slope: Treat as a flat roof. Use a continuous membrane system: either built up felt/asphalt with crushed stone or metal system with sealed or soldered seams.	

FIGURE 13.31 Roofing materials lowest permissible slope.

Traditional	Metric
2/12	50/300
4/12	100/300
6/12	150/300
8/12	200/300
10/12	250/300
12/12	300/300

FIGURE 13.32 Roof pitches.

develop problems in extremely cold temperatures. I have found this to be true in Maine. My experience has been mirrored with other contractors. Local suppliers have commented to me on the number of warranty calls they receive on fiberglass shingles. My talks with local contractors and suppliers have proved out what my experience has shown. Fiberglass shingles can give contractors problems in cold weather.

Fiberglass shingles installed in my area have a history of cracking and blowing off. I assume this is due to the cold temperature. However, I cannot say with certainty that temperature is what causes the shingles to crack. The professionals I have spoken with concur with my opinion that cold temperature is the primary cause for fiberglass shingles failing. Two of the big selling features of fiberglass shingles are their reported insulating qualities and fire-resistance.

Wood

Wood shingles and shakes are a viable roof for some types of homes. These roofs are, however, something of a fire risk unless they are treated to increase their fire-resistance. Wood roofs are expensive, and they are not normally used in most modern construction. This doesn't mean that you won't have occasions to work with wood. Some existing homes will have wood-covered roofs. As you go about your remodeling work, you will find times when you have to work with wood shingles and shakes.

Tile

Tile roofs are not common in most regions. This doesn't mean that there are not thousands of tile roofs out there, just waiting for a remodeling contractor. Moving around on a tile roof is tricky. The tiles can be slippery, and they can break easily. Few customers will have tile roofs or request a tile roof, but you might find a job here and there where tile is used.

Slate

Slate was a popular roofing material years ago. While I have not had much experience with tile roofs, I have worked with numerous slate roofs. I can tell you from experience that slate roofs can be very slippery. It is not unusual for moss-type vegetation to grow on a slate roof. This make footing even more treacherous. Slate is also very heavy, and brittle. Working with slate effectively requires experience, caution, and patience.

ATTENTION TO DETAIL

If you pay attention to detail, roofing can be easy to understand. For example, plywood applied to trusses on 2-foot centers should be connected with plywood clips (Figures 13.33) to prevent sagging. Whether you are working with

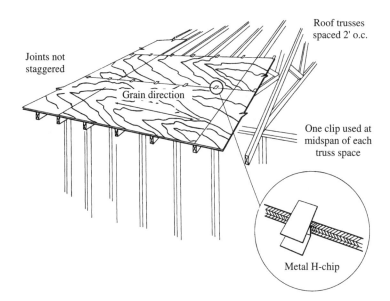

FIGURE 13.33 Plywood clip installation. (courtesy U.S. government)

a gable roof, a hip roof, or a roof with dormers (Figure 13.34), you should have a clear detail of what your framing plan (Figures 13.35 and 13.36) is. All things considered, roofing is not complicated to understand, but it does require planning to be done well.

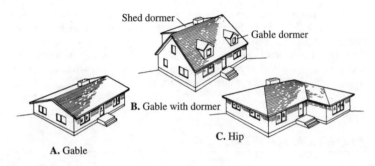

FIGURE 13.34   Roof pitch types. (courtesy U.S. government)

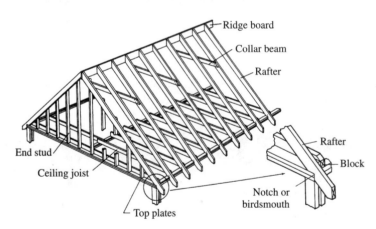

FIGURE 13.35   Typical rafter framing. (courtesy U.S. government)

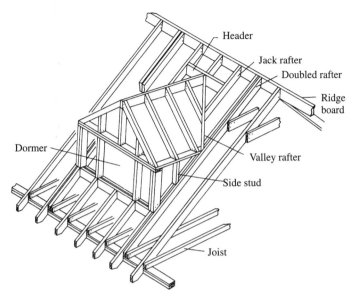

FIGURE 13.36 Gable dormer framing. (courtesy U.S. government)

CHAPTER 14
STEPS, STAIRS, AND RAILINGS

Stairs are often taken for granted. People assume that stairs can be installed where they are needed whenever they are needed. This might be true, but functional stairs are not always feasible in all locations. While it is rare when remodelers have to alter an existing set of stairs, occasions do occur when this work is required.

For example, if a person is finishing off a basement, new stairs might be needed. These could be replacements for old basement stairs, or they could be the first set of stairs ever installed to allow access from the main living area into the basement. The same set of circumstances could apply to an attic conversion. So, there are times when providing new stairs in a remodeling effort is necessary. Of course, new construction also requires the building of steps and stairs.

Building a set of rough steps is not a tough job, but designing an efficient layout for a difficult situation requires skill. Designing a set of stairs is often much more difficult than the actual construction. Will a landing be needed? Should the stairs have a winder in them? Could a spiral staircase be used? These are just a few of the questions that might come to mind when factoring a set of stairs into a job.

Building codes regulate certain dimensions pertaining to stairs. For example, the tread width and the riser height is normally regulated. So is the amount of headroom and the width of the stairs. Provisions for handrails are also normally dealt with in local building codes. Whatever the local building requirements are, you must be aware of them and adhere to them. This can complicate your job at times.

Local jurisdictions are empowered to modify the buildings codes that they adopt. This means that rules for local codes might not be identical to those found in general code books. I can't tell you what the requirements for stairs are in your area, but I can give you some rule-of-thumb figures that should be close to what your local code mandates.

Primary stairs, meaning most stairs in living space, should have a minimum width of 32 inches, and 36 inches is better. There should be a minimum of 30 inches of open space when measuring between the handrails or the handrail and the wall. A maximum riser height should not exceed 7½ inches. All treads should have a minimum width of 9 inches. Minimum head clearance should be set at 80 inches. When a landing is installed, it should have a minimum depth of 36 inches. Handrails are usually installed so that they are between 30 and 33 inches above the walking surface. You can normally get by with just one handrail with a closed staircase, but some areas might require two.

Secondary steps are often subject to less stringent code requirements. However, this is not always the case, so check with your local building inspector. When secondary steps are identified, they are usually steps going into an unfinished basement or attic. While these steps might not have to conform to requirements for primary stairs when they are used with unfinished space, remodeling the space into living space can trigger a rule where the stairs must be upgraded to meet code requirements for primary stairs. Again, these numbers are not necessarily what is required by your local code, so check on the local level before planning or building stairs.

EXTERIOR STAIRS

Exterior stairs or steps are often a part of building projects. If you build a deck for a customer, you will probably have to build steps for it (Figure 14.1). The same could apply to the construction of a screened porch. Complex designs are not normally incorporated into exterior stairs. The most complicated design typically encountered involves a platform so that the stair can offset at a 90-degree angle. There is no real challenge here. Stringers come off the upper landing and go straight to the platform. From there, another set of steps descends straight down to the ground. Some outside construction might get a little more involved, but most outdoor stairs are simple to figure.

In addition to decks and porches, outside steps might be used to access a room over a garage. Some houses might have exterior steps leading up to space created in an attic conversion. Computing the rise and location of these more extensive stairs can become more difficult. However, the same basic principals used to calculate interior stairs apply when working with exterior stairs. Because most work with stairs will be done inside a home, let's concentrate our time on them.

INTERIOR STAIRS

Interior stairs can be built in a number of configurations (Figure 14.2). Most, however, are either straight runs or have a winder or platform in them. More exotic versions might be built with a sweeping arc. Some stairs are spiral in design and require minimum space for installation. There are open-tread stairs

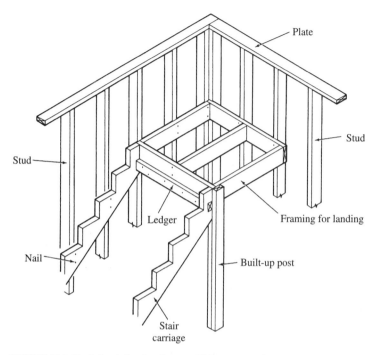

FIGURE 14.1 Typical stair framing. (courtesy U.S. government)

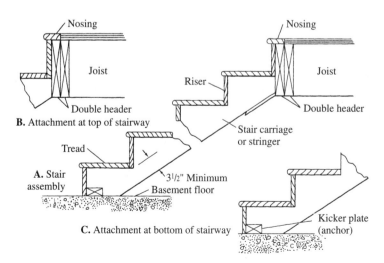

FIGURE 14.2 Attachment at bottom of stairway. (courtesy U.S. government)

and closed-tread stairs. Some stairways are enclosed with walls on each side. Others have a wall on one side and a railing on the other. A few have railings on both sides and no walls. Each type of stairway has it good points, but they all have their bad points, too. It will be up to you and your customers to figure out what type of stairs will work best.

A Basement Conversion

A basement conversion can require you to upgrade existing basement steps or to build a complete set of stairs. If you are upgrading an existing set of stairs, the work will be self-explanatory, once you are sure of your local code requirements (Figure 14.3). You might have to add an extra stringer. There will probably be a need for additional railing and pickets.

It might be that the rise or headroom cannot meet code requirements for a primary stairway. In this case, you might as well start from scratch. In fact, I have found that it is usually prudent to remove rickety old basement stairs and build new ones right from the start. If customers are going to spend major money to finish off a basement, they might as well have decent stairs leading into the space. Besides, it can take longer to tinker with old stairs than it would to build new ones.

I have seen many jobs where the existing basement stairs were in the way. Their placement made it impossible to get maximum utilization out of available space. Sometimes the stairs have been in the only feasible place they could be, but I've seen many jobs where the basement steps could be relocated. Whether you are moving the stair location completely or modifying the way in which the stairs are installed, you can enhance some basement conversions with this type of work. Not only will this make the customer happy, if you play up your ideas during the bidding and selling process, your input on the stairs could be enough to win you the job. Let me expand on this for a moment.

Let's say that you've been called out to give an estimate on a basement conversion. The homeowners escort you into the basement and point their plans out to you. As you look around, you see that the existing stairs are built so that the end of them restricts the size of a new family room the couple wants. Other contractors have already been on the job, and they took the customer's recommendations at face value, never mentioning other options. You, on the other hand, speak up with ideas on how the basement could be more

- Minimum stair width is 32 inches.
- Minimum nosing with closed risers is 1 inch.
- Minimum overhead clearance is 80 inches.
- Preferred overhead clearance is 88 inches or more.
- Preferred railing height is 36 inches.

FIGURE 14.3 General residential stair requirements.

spacious with a different stair location. Well, the homeowners look at you with big eyes and wide smiles. You are the first person to show them a way to get a bigger family room, which happens to be their main reason for converting the basement space. By using your knowledge of stair designs, you have just gained a competitive advantage over all the other contractors. Little sales maneuvers like this can boost your sales and your income.

When you are looking at a basement conversion, you should investigate all options for the stairs leading into the space. If existing stairs are in place, obviously it will be easier to use that same location. However, easier is not always better. If the stairs come down in the middle of the basement, your customer might prefer to have new steps built that could come down a side wall and not take up so much valuable space. Even just adding a landing and a 90-degree turn near the bottom of the steps could open up enough room for a hallway. It is always best to approach remodeling with an open mind. The more you study the circumstances, the more likely you are to find better ways of accomplishing your goal.

Attic Conversions

Attic conversions require the building of new steps. In some cases, there might be a set of secondary steps installed to the attic for access, but these steps will generally have to be upgraded. A lot of attics have only scuttle holes or pull-down stairs for access. Under these conditions, you must find a place for a full-size set of stairs. This can be a bit of a challenge. It is fairly easy to work in a set of access stairs; however, to install formal stairs that are accessible from public areas is a whole other matter.

When faced with tight quarters for installing a conventional stairway, many remodelers turn to spiral stairs. These units do take up a minimum of room, but they also have some drawbacks. Spiral stairs are difficult for some people to climb. The shape and size of a spiral stairway makes it very difficult, if not impossible, to move furniture up and down them. The seemingly vertical climb up a spiral stairway is another strike against this type of unit.

I'm not suggesting that you should never use spiral units, but you and your customer had better think through all angles of the proposition before proceeding. In fact, let me tell you a couple of quick stories about two different jobs where spiral stairs were on the list of things to do.

The first job I'm going to tell you about involved a friend of mine. He and his wife had a new home built for them many years ago. I wasn't building houses at that time; I was concentrating on remodeling. Anyway, the house was built with a full, walk-out basement. My friend wanted a spiral stairway installed between the basement and his living room. One of his reasons for wanting this was so that he could carry wood up from the basement for the stove in his living room. Another friend of mine was the carpenter on this job. He built the house and the stairs to the specifications provided by the owner. Everything went well through the summer months. Well, almost everything.

The new house had laundry facilities in the basement. The new home-owners soon discovered that toting laundry baskets up and down the spiral stairs was no fun. The narrow stairs made it difficult to navigate with a basket full of laundry, and the steep climb was tiring on the way up. As if this was not bad enough, things got worse in the winter.

A month or so into the heating season, my friend called and asked me to come over at my first opportunity. He didn't say anything about his stairs or about anything else that was related to work. I assumed I was being invited over for a social call. The following Saturday found me sitting in my friend's living room. It was toasty warm and there was a substantial pile of wood near the stove.

As I admired my friend's new home, I couldn't help noticing pieces of bark and wood debris between the wood pile and the sliding-glass door from the deck. This seemed strange to me, because the wood was stored in the basement. Why would anyone go out onto an icy deck, down exterior steps, and lug wood up from a basement when they could use interior stairs? I questioned my buddy on this.

I was told that the spiral stairs were just too difficult to climb with an arm-load of wood. The conversation continued and revolved around the spiral stairs. It turned out that one of the main reasons I had been invited out was to give some free professional advice on what other stair options were available. We talked for a good while, and I offered some suggestions. A few weeks later, my crew was working on replacing the spiral stairs with a conventional stairway. My friend threw away a good deal of money by not thinking his plan through all the way in the first place.

Moving onto another job, I can tell you about a customer who was saved from the type of pain described in the last example. This person wanted to convert the attic over his garage into office space. His idea was to use a spiral stairway to gain access to the office. He thought it would look nice for his clients, and he liked the fact that the unit wouldn't deprive him of much space in his garage. When I arrived on the job to give an estimate, the stairs were quickly a topic of conversation.

As we talked, I learned of the homeowner's plan to use the new space for an office. I inquired as to whether there would be desks in the office. He looked at me with a strange stare, and confirmed that there would be desks in the office. I stopped him right there by asking how he planned to get the desks into the office. It was as if I had hit the man with a baseball bat. His face went blank. He looked up at the garage ceiling and then back at me. After rustling through some papers in a folder he was holding, he produced specifications for the spiral stairs he had almost ordered. It only took a few moments for the man to realize how badly he would have felt if the spiral stairs had been installed. You see, the stairs he had in mind were much to small to allow free passage of a desk.

In my opinion, spiral stairs should be used for decorative effects and secondary steps, but not at primary steps. They are fine if you just want occasional access to a room that will not contain large furniture; however, if the stairs will see daily use, then spiral units are probably not the best choice.

OPEN OR CLOSED

When you are consulting homeowners on the types of stairs to have installed, you will have to address the issue of whether the stairs should be open or closed. Of course, existing conditions will have something to do with the final decision. There are advantages and disadvantages to both types of stairs.

Closed stairs are, in my opinion, a little safer when used by children and elderly persons. Kids love to slide down banisters, and this can be a dangerous pastime. Children also try to make open stairs into their own personal playground. Balusters are not monkey bars, and they can break away, allowing a child to fall. If an adult slips on the stairs, and this happens, the momentum of their fall could cause the railing and balusters to break. This might result in a much more nasty fall. For these reasons, I prefer a closed stairway under certain circumstances.

Closed stairs are not as attractive as open stairs. They don't allow light to flood the stairs, and the closed-in effect can be confining. It is also true that an open stairway tends to make a home appear larger. Anytime an open concept is used, rooms seem to look bigger. From an image point of view, I prefer an open stairway. Yet, I like the functional qualities of a closed staircase. Because much of the debate over open and closed stairs hinges on personal preference, the outcome is subject to an individual's own preferences.

PLANNING

Planning is the key to success when it comes to stairs (Figure 14.4). You should explore all options before making a firm commitment on how to build a particular set of stairs.

For example, if your customers are advancing in age, it might be wise to build their stairs with multiple landings installed. This will give them a place

---

- Divide total floor to floor rise (in inches) by 7.
- Move fractions to nearest whole number.
- This is the number of risers that you will need.
- If you divide the total rise by the number of risers, you can determine the height of the risers.
- The number of treads will be one less than the number of risers.
- Divide the total run, in inches, by the number of treads to determine the length of treads.
- Treads and risers must be in proper proportion to pass local code requirements.

FIGURE 14.4 Steps for calculating stairs.

to stop and rest along their climb. You might have an occasion when an oversized width would make negotiating stairs easier for your customer.

Every job will have its own set of circumstances to work around. It's your job to keep trying out different ideas on paper until you arrive at an ideal solution for your stair installation.

CHAPTER 15
WINDOWS
AND SKYLIGHTS

Windows and skylights are the eyes of a home. Many jobs involve the installation of new windows or the replacement of existing windows. Almost any home can benefit from the addition of windows and skylights. The more natural light a house has, the bigger it looks. Also, bright rooms are more appealing rooms to be in. This, along with other reasons, is what makes the use of new and replacement windows so popular.

Windows are such big business in the remodeling field that there are contractors who specialize in them. Some contractors spend nearly all of their time working with windows. If your company is not getting a big piece of this action, you are missing out on some money. However, whether you specialize in replacement windows or just work with windows as part of your routine building, there is money to be made from them. To capture some of this capital, you have to know how to sell and install windows in a way that will make your customers happy.

CONFUSION

Confusion runs high with consumers who try to sort out their window options (Figure 15.1) without the expert guidance of a professional contractor. Sales associates in supply houses inundate consumers with literature and sales hype. Much of this only serves to make matters worse for the average person. The confusion starts with U-factors and runs a long gamut. You can take advantage of this situation. If you can assist your customers in sorting out which windows will be best for them, you are more likely to win their confidence and jobs.

For you to be able to consult efficiently with your customers, you have to have a depth of knowledge in the subject being covered. In this case, we are talking about windows and skylights. If you were given a test on windows, do

- Double-hung
- Casement
- Fixed
- Awnings
- Sliding
- Skylight
- Bay
- Bow

FIGURE 15.1 Types of residential windows.

you think you pass with flying colors? If you take windows and skylights for granted, you are shortchanging your income. There is enough money to be made in windows to warrant a little extra time being invested to upgrade your knowledge. With this in mind, let's begin our lesson on windows and sky-lights.

TYPES OF WINDOWS

How many types of windows are there? A lot. There are more than enough variations in windows to make anyone happy. Almost everyone thinks of a double-hung window when asked to visualize a window. Is this what you would think of?

Hey, let's play a little game. Take a moment to think of six types of windows. Don't take too long on this; you should be able to name at least six or eight types of windows right off the top of your head. Did you come up with six different types of windows? Well, let's compare your answers with the following descriptions of different types of windows. If you didn't do well on this exercise, you need to hone your knowledge of windows.

Double-hung Windows

Double-hung windows are probably the most well-known windows in use. They are quite common and very popular. I would guess that there are more double-hung windows in use than all other types of windows combined.

What makes these windows so favorable? Part of their popularity is cost. Double-hung windows are affordable, in terms of window prices. Tradition probably plays a part in the choice of double-hung windows. I suspect that the main reason double-hung windows are so prolific is that they are the type of window that most contractors push on their customers. This is not said to de-mean contractors. My point is that few contractors offer their customers a full range of window choices.

Single-hung Windows

Single-hung windows look like double-hung windows. The difference between the two is that both sashes move in a double-hung window and only one sash, usually the bottom one, moves with a single-hung window. Single-hung windows are a little less expensive than double-hung windows, but I don't think the savings is enough to offset the popularity of double-hung windows.

There is something ironic about the comparison of double- and single-hung windows. Many people shun single-hung windows as being cheap products. As I've already said, the primary difference between single- and double-hung windows is that only one sash moves with a single-hung window. Now, a double-hung window can have only one sash fully open at a time. There is more control available when double-hung windows are used. The lower sash can be raised, and the upper sash can be lowered. I know there must be people who use this option to their advantage, but most people always raise the lower sash. If this is going to be the case, why not just use single-hung windows. I'm not saying that you should convince your customers to go with single-hung windows, but it does seem that with the habits of most homeowners, single-hung windows would work just as well as double-hung windows.

Casement Windows

Casement windows are more expensive than single- and double-hung windows. However, they also offer advantages that the other windows can't. For example, a double-hung window can never have more than half of its total size open. This is not true with a casement window.

Because casement windows are hinged on the side and crank out, it is possible to get full air flow through the entire size of a casement unit. This can be a big advantage. Additionally, casement windows are typically more energy efficient than double-hung windows. While casement windows cost more to buy, they can save their purchaser money in utility bills over future years. This savings can more than pay for the extra acquisition cost.

There is another big advantage to casement windows that a lot of people never think about. How many times have you struggled to push open a double-hung window? Can you imagine how difficult raising a double-hung sash is for people with limited strength or physical disadvantages? Think about it. Opening a double-hung window can be a major chore for some people. When compared with how much easier it is to crank open a casement window, double-hung windows might not be suitable for elderly residents, children, or people with physical limitations. Keep this in mind when talking with your customers. Because good windows are a long-term investment, you will do well to look into the future for customers who might grow into a stage where operating double-hung windows will be a struggle.

Awning Windows

I'll bet you didn't think of awning windows when you were compiling your mental list. These windows are what I consider special-purpose windows. They are not the type of window that would normally be installed throughout a house. Awning windows do, however, have their strengths.

Because awning windows are hinged at the top and open upward, they can be left open in many rains without having water come through the window opening. A blowing rain can still invade living space; however, if the precipitation is falling straight down, an awning window will allow ventilation while blocking out the foul weather.

Privacy can be maintained when awning windows are used. These windows can be installed high on a wall and still be opened easily. This provides light and ventilation at a height that is high enough not to encroach on someone's privacy, such as in a bathroom. Another good place for awning windows is in a sun room. If you are remodeling a room to be used for a spa enclosure, awning windows will allow moisture from the spa to escape while maintaining privacy. These windows should be used sparingly, but there are definitely times when they should be used.

Sliding Windows

Sliding windows were popular years ago, but they are not in such good standing today. Sliders are typically considered to be cheap windows that don't offer good energy efficiency. While it is still possible to find some sliding windows being installed, they don't rate high on the list of desirable windows for many people.

Bay and Bow Windows

Bay and bow windows can change both the interior and exterior appearance of a home. It is also possible to increase the square footage of living space with a bay window. You must, however, take into consider the additional cost of foundation work, framing, and finish work that goes into the building of a bay window. Nail-on bay windows brighten up a room dramatically, and they enhance the overall appearance of a house.

Both of these window units run on the expensive side, but there are times when their cost is justified. One example of this might be when remodeling an eat-in-kitchen. The construction and installation of a bay window can create a nice nook for a breakfast table. This isn't a cheap proposition, but on the other hand, it will not cost a fortune.

Garden Windows

Garden windows can be used anywhere, but they are particularly popular in kitchens. These expansive windows let in a lot of light, and their design allows plants or other items to be set in them. Garden windows come in various styles, so you will have to shop around to see all of the features you can offer your customers.

Fixed-glass Panels

Fixed-glass panels don't show up as often as I think they should. I've used fixed glass in countless jobs. It is an inexpensive way to let huge amounts of light flood into a home. The low cost of fixed-glass panels make them desirable in sun rooms and other rooms where ventilation is not your goal. These panels can also be used in conjunction with operable windows. By doing this, you get a wall of windows for a much lower cost. Any good glass company can provide you with a wide variety of glass panels.

SKYLIGHTS

Skylights provide tremendous opportunities for remodelers and building contractors. If you are fighting a dark room, skylights can be the answer. Of course, you must have a direct path to a roof before you can make use of a skylight. It is not mandatory that you have a vaulted ceiling to work with. If there is attic space over the location where you wish to install a skylight, you can frame in a light box. Just as there are a number of window choices available, so are there a lot of skylights to work with.

Plastic Bubbles

Plastic bubbles are the least expensive type of skylights. These domes come in various shapes to accommodate nearly every need. Some of these units have a flange that is installed directly under roof shingles. Other models are made to be curb mounted. Both types are pretty easy to install, and the units themselves are very inexpensive.

There are drawbacks to these low-cost skylights. One of the biggest disadvantages is the lack of energy efficiency. Being made of plastic-type material, the insulating quality of a bubble skylight is poor. Not only do these skylights cause heat loss, they tend to condensate badly at times. This factor alone can be enough to disqualify the units from consideration. When skylights condensate, the light boxes that accommodate the units suffer from water stains. Flooring below the skylight can become damaged from dripping water. If your customers insist on using uninsulated skylights of any type, be sure to protect yourself from upcoming complaints of condensation.

If your customers are seeking skylights to help exhaust summer heat from their home, plastic bubbles are not going to get the job done. These units are fixed, they do not open. Skylights can be very effective in venting summer heat from a house, but these fixed units are no good for this purpose.

Plastic bubbles are available in clear colors and in tinted colors. I have always favored a bronze tint when using these skylights, but that is mostly a personal preference. The installation of skylights can cause carpeting and furniture to fade, due to direct sunlight. This is something you should point out to your customers. Clear skylights can cause the fading process to accelerate.

Fixed Glass

Fixed-glass skylights come in various degrees of quality. These skylights also come in many different sizes. Skylights with stationary glass panels can range from the very cheap to the very expensive. Their quality can rival that of any other skylight. The only major drawback to skylights with fixed glass is their inability to provide any ventilation or exhaust potential.

Operable Units

Operable units offer the most flexibility in the selection of skylights. These units can be opened to remove summer heat, and they can catch some breezes when mounted in steep roof pitches. I just finished building a new house for myself, and I installed an operable skylight in my office. It is equipped with a screen and is designed to be opened with a pole. This can be done manually or with an automatic opener. In my particular case, the skylight is mounted low enough in the roof that I can reach it and open it by hand. I should also say that I installed two skylights in my great room. These are fixed units. Operable units are very nice, but they are also expensive.

Roof Windows

Roof windows provide a reasonable alternative to building small dormers where the only real purpose of the dormer is to allow the installation of a window. Roof windows can let in more light than a window in a dormer, and they eliminate the need for such extensive framing.

If I were remodeling a Cape-style house for myself, I might use dormers to give the building a traditional architectural look, but otherwise, I would rely on roof windows for my light and ventilation. Roof windows are available with screens, built-in blinds, and other options. These units are expensive, but they are a cost-effective alternative to building dormers.

INSTALLING WINDOWS

Installing windows is not normally a difficult task (Figure 15.2). There are times, however, when unexpected circumstances can throw you a curve. For ex-

- The tops of windows should be set at 80 inches.
- Sill heights for windows adjacent to counters should be at least 42 inches.
- Sill heights for windows near furniture should be at least 42 inches.
- Sill heights for picture windows should not exceed 38 inches.

FIGURE 15.2 Suggested window measurements.

ample, a customer might want to replace an existing kitchen window with a new garden window. Because the garden window will probably be larger than the existing window, you must plan of framing a new rough opening. If you've been in the remodeling business for long, this part of the job won't present any problems for you. However, what happens if the plumbing vent for the kitchen sink is installed in the stud bay next to the existing window. The new garden window will need to occupy this space, but there is a pipe in it. What would you do?

If a vent pipe is running up in an area that will interfere with a window installation, you must have it relocated. This usually isn't a big job, but plumber's aren't cheap, and even small jobs can have big price tags. A plumber will probably offset the vent into the next clear stud bay and then tie back into the existing vent before it exits the roof.

However, it might be necessary to invest more time in the job. Your plumber might have to get under the kitchen sink and reroute the vent, and drain, from below the floor level. This will run your cost up somewhat.

Relocating the plumbing won't take long and even if you hadn't counted on the expense, you won't have to hock your truck to pay for it. However, it is always best to avoid unexpected expenses. In this case, you could do that with a clause in your contract that protects you if unforeseen obstacles are found within the wall.

Electrical wires might very well present some trouble for you when installing new windows. Most good electricians run their wires low enough or high enough so that they will not interfere with future windows, but not all electricians are so thoughtful. You could open up a wall and find a string of wires running right through the proposed window location. The wires can be moved, but again, you are incurring additional expense. This is why it is good business to have certain clauses in your contract to give you a way to recover your expense in dealing with unexpected costs.

Check and Double Check

Check and double check the required rough-opening measurements before you start cutting into the siding on a customer's home (Figures 15.3 and 15.4). There are few things more embarrassing that cutting a hole in the side of someone's home that is too big for the new window being installed. It's easy to make holes larger, but it's tough to shrink them.

Wind velocity (mph)	Window glass thickness (in.) @ 0.133
30	64.5 square feet
40	32.25 square feet
65	16.1 square feet
120	4.5 square feet

FIGURE 15.3 Maximum range of glass size, based on wind velocity.

Wind velocity (mph)	Window glass thickness (in.) @ 0.085
30	30 square feet
40	17.5 square feet
65	8.5 square feet
120	2.5 square feet

FIGURE 15.4 Maximum range of glass size, based on wind velocity.

When you are measuring for rough-openings, don't take anything for granted. I've seen carpenters measure one or two windows and assume that all the remaining windows were the same. People don't always choose windows with identical measurements, so an assumption like this can put you in hot water. You should measure and research each individual window before you take any action for it.

Look on Both Sides

Before you start cutting in a new window, look on both sides of the wall where the window will be installed. I know this sounds stupid, because any professional remodeler should have the common sense to do this, but not all of them do. Walking around the interior of a home with the homeowner showing you where windows are desired can distract you from outside obstacles.

If you fail to think ahead and investigate what's on the other side of the wall, all kinds of things can happen. At the least, you could have an irate customer on your hands when you inform the homeowner that a window placed in an agreed-upon location will not work. Worse yet, you might even start cutting in the window, from the inside, without realizing that you are going to run into a roadblock. Taking a walk around the exterior of the home will allow you to avoid this problem.

What types of outside obstacles might get in the way of a new window? A bulkhead door can be a problem, and so can an attached tool shed. You'd feel pretty funny cutting a new window into a dining room only to get a view of the inside of a tool shed, wouldn't you? If a home depends on fuel oil for its heating system, the fill and vent pipes for the oil tank, or even the oil tank itself, could block the installation of a window. All of these situations are easy to spot if you look for them, but you have to take the first step and look for them.

Watch Out for Hidden Masonry Walls

When you are quoting a price to install new windows, you should watch out for hidden masonry walls. I had a friend, who is a remodeler, run into this problem just a few months ago. The contractor took on a major remodeling job. He was converting a building from two-family use into four-family use. The job required extensive interior work, and the property owner wanted all new windows. In addition to replacing existing windows, a lot of new windows were to be added due to the expansion of rental units. My friend has a lot of experience in remodeling, but he goofed on this job. When he gave the property owner a price for the work, he had no idea of what was waiting for him inside the exterior walls.

It turned out that the building being remodeled was built long ago, when attacks from hostile groups were not uncommon. As protection from bullets and arrows, the original builders constructed the exterior walls with brick, and a lot of it. Over the course of time, various remodeling efforts had been made on the property, and it was covered with wood siding on the exterior and plaster on the interior by the time my buddy got involved with it. He saw the clapboard siding and the plaster and assumed he was bidding a typical old job. However, when he went to cut in the first window, he hit a brick wall, quite literally.

I went to the job site to meet my friend for lunch one day, and I saw the brick work. The number of windows being installed escapes me at the moment, but I know there were a lot of them. Can you imagine having your job go from a simple one to this type of nightmare? I can't even begin to guess how much money my friend could have lost on the job. Fortunately, the property owner had deep pockets and a kind heart. Rather than hold my friend to his written quote, the property owner agreed to pay extra for the window installation.

In 20 years of remodeling, this was the first job I had ever seen where brick had been covered up with wood siding, but I promise you, it was. This just goes to show, even veteran remodelers can run into big problems without ever seeing them coming. However, a good contract that excludes liability for such hidden expenses can help to protect you.

WORKING WITH SKYLIGHTS

Working with skylights can lead to some interesting experiences. Most skylight installations go off without a hitch, but there are times when existing conditions make the work something of an adventure. Any time you are working with old houses, you can run into strange and unusual circumstances. For example, you might find that the roof structure is not made of typical rafters or trusses. It could be made up of old beams. If this is the case, you are going to have some extra framing work to do before you can install a skylight.

Roofs on older homes can look pretty good and be in horrible shape. If you don't get up on the roof before you quote a price for installing a skylight, you could be making a big mistake.

Suppose the existing roof is too brittle to work with? What will you do if the house has a slate roof and you have failed to notice this fact until after you have committed to a price for installing two skylights? Have you ever considered that the front roof on a house could be protected with asphalt shingles while tin covered the bathroom roof? It can happen. You can't afford to take anything for granted when you are a remodeling contractor.

Hey, I've even seen houses where parts of the roof were covered with old metal signs. Yes, signs. In fact, there is a house, here in Maine, not more than 30 miles from where I live that has its entire roof covered with old signs. How do you flash a skylight into a rusted old metal sign?

Skylight installations can be easy or hard. The type of roof covering you are working with certainly plays a part in which way the job will go. The rafter configuration is another element to be considered. In addition, the condition of an attic can affect the speed with which you can install a skylight. Insulation can get in your way. Puny bottom cords can make walking around in the attic treacherous. You have to investigate all existing conditions carefully before you start throwing figures around.

Once you have covered all the bases where problems might pop up from existing conditions, you can go about your work in an efficient way. The actual installation of a skylight, under normal conditions, is not difficult. Jobs that require long light boxes get a little tricky, but they don't offer much trouble to an experienced carpenter. Flashing the skylights properly is one of the most important parts of an installation. As long as you follow the installation instructions provided by the manufacturer, you are not likely to encounter any problems. However, some contractors think they know it all and don't need instructions. This is silly. Not all skylights are alike. If you fail to read and follow the instructions packed with a skylight, you could be setting yourself up for trouble down the road.

HELPING YOUR CUSTOMERS

Helping your customers determine what type and style of windows they should buy is a big responsibility. So is helping them sift through sales hype and technical jargon, but it is a part of your job. If you don't handle this side of your job well, you might not have any work to do. Let's spend the remainder of this chapter discussing the many ways for you to assist your customers in making decisions on windows and skylights.

Light

Additional light is always one benefit derived from adding new windows and skylights. Surveys have shown that home buyers like rooms that are bright and cheerful; this makes adding natural light to a room a good investment, within reason. You can point out to your customers what rooms in their home will benefit the most from various types of windows and skylights. For example, a skylight installed over a new whirlpool tub can provide some nice evenings of star gazing while relaxing. Skylights in a kitchen can brighten the work area and make the room more comfortable to be in. A garden window can also add more light to a kitchen.

Appearance

Appearance is always important when considering home improvements. Adding a new window can balance the exterior elevation of a home. Architectural enhancements can come in the form of bay and bow windows. Windows with arched transoms above them can set a house apart from the rest of the homes in a given area. When you are thinking about, and selling, new windows, you have to consider all of your options.

Necessity

Some windows are installed out of necessity. For example, if you are creating a new bedroom, you might have to install a new window to meet egress requirements in the building code. Attic conversions typically require the installation of new windows, and roof windows are a good option in these cases. If a customer is being forced to add new windows, you should offer a wide selection of options for consideration.

Ventilation

Ventilation is, of course, a primary reason for having windows in a home. When this is the impetus behind installing new windows, you should describe how different windows work in a way that your customer will understand. For example, you should point out that casement windows allow full air flow while double-hung windows are restricted to one-half of their size in air flow. Awning windows also give complete air flow, so lay out all of the options and explain them to your customers.

Skylights can also be used as ventilation tools, assuming that they are capable of being opened. A house equipped with high ceilings, reversible-rotation ceiling fans and operable skylights can provide many ventilation options.

Floor Space

If only a little floor space is needed to tweak a room into prime condition, the construction of a bay window might be just the ticket. You can control the size of the unit and the floor space achieved. It might be wise to use a combination of fixed glass and operable panels to make a cost-effective bay window.

Sun Rooms

Sun rooms are very popular additions to homes. When I was working in Virginia, it seemed that we were constantly building sun rooms. Casement windows, fixed glass, and awning windows all work well together in this type of remodeling project. A lot of sun rooms are done with sliding-glass doors, but ventilation is limited when this approach is taken. The use of casement and awning windows eliminates the restrictions associated with sliders, whether they be sliding doors or sliding windows.

Energy Efficiency

Energy efficiency has become a big issue with windows. Most homeowners don't know how to judge the efficiency of a window they are considering. One factor is, of course, the U-factor of the glass. You can point out to home-owner's how to assess the insulating quality of glass by its U-factor.

Most homeowners are familiar with R-factors, but many have never even heard of U-factors. Because U-factor ratings work in basically the opposite direction of R-factors, many inexperienced people buy poor quality windows, thinking that they have done well. When discussing energy efficiency, you must educate your customers on window design and construction. Show them why a casement window is usually a better value, in terms of energy efficiency, than a double-hung window.

I recommend taking cut-away sections of various windows to the job with you during the bidding process. It is much easier for people to understand information you are providing if they can see examples of what you are talking about.

Maintenance

Maintenance is another factor to be dealt with when talking about windows. Does your customer want a clad window that requires minimum maintenance or a wood-frame window that is cheaper but needs routine painting? If you are not confident of your knowledge in this, or any other, area of windows, talk to your suppliers. Manufacturers will be more than happy to supply you with sales brochures and technical information. When you are in a selling mode, you product knowledge might be all that is needed to take a job away from a lower bidder.

Organize Your Material

Organize your material well. A solid sales presentation always benefits from a well organized proposal. Good organization of your material will also make it easier for your customers to follow you through your presentation. If you are going to talk about Low-E glass or Argon gas, have supporting documentation available to hand out to your customers. This will not only give them something tangible to assess, it will get them involved in your presentation. This is a key to making sales.

CHAPTER 16
HVAC SYSTEMS

Heating and air conditioning are often standard equipment in homes today. While there are still houses where air conditioning is not needed or is not present, there are very few homes without heat in them. Most houses have central heating systems, though some are still being heated by only wood stoves and space heaters. Heating and air conditioning, for the most part, are considered to be mandatory equipment in a home.

As a general contractor, you are not likely to get called to replace a furnace or to install a new air-conditioning system. You are, however, very likely to get caught up in this type of work when it is being done in conjunction with remodeling or the construction of a new home. Most contractors bear some burden of responsibility in the area of heating and air conditioning. Are you prepared to accept this role? If not, you need to hone your skills in talking intelligently about heat pumps, forced-air furnaces, hot-water heating units, and so forth.

EVALUATING EXISTING SYSTEMS

If you are a remodeler, you are going to be forced into working with heating and air conditioning, and you are going to find yourself evaluating existing systems. Unless you have a lot of experience in this field, you should take an expert with you when inspecting existing units. A boiler that looks perfectly fine might have a cracked section. The heat exchanger on a heating system can be bad and very difficult for untrained eyes to detect. If you are adding space to a house, you will have to determine if existing heating and cooling systems can be tapped into. They might not be large enough to handle the increased demand of extra living space. If you forget to factor in a new heating system when doing your cost estimates, you might lose a lot of sleep trying to figure out how to worm your way out of the mistake. It is definitely in your best interest to take a professional along with you on routine inspections of existing systems.

Some contractors will say just about anything to get work. Others are just ignorant enough to make statements that put qualified contractors up against a wall. Let me give you an example of what I'm talking about. Let's say that you are estimating an attic conversion. As part of your estimate, you are concerned about getting heat to the upstairs living space. You're not comfortable that the forced hot-air furnace and existing duct work will be adequate. On top of this, the customer wants you to install a new air-conditioning unit and tie it into the existing duct work that is used by the furnace. After listening to the customer, you express your concerns about the existing equipment not being suitable for the job. You recommend installing a one-piece heat pump in the attic, because it will be all one big room, used as a studio.

After making your feelings known, the customer replies by saying that two other contractors didn't have any problem with doing the job the way it was specified. The homeowner starts to question your knowledge and ability. In reality, you are right and the other contractors are wrong, but the homeowner is choosing to side with the previous contractors. This might be because two independent contractors didn't raise the same concerns that you have, or it could be that the customer doesn't like to be wrong. Either way, you're in the hot seat.

Now that you've got your potential customer questioning your experience, what are you going to do? If you're right, and in this case you are, you should stand your ground. This might cost you a sale, but it is better to lose the sale than to take it and wind up in a big mess. If you had an expert with you who knew heating and air conditioning as well as you know general remodeling, you could sway the customer. While it's true that two remodelers didn't voice concerns over the mechanical equipment, having an expert back you up could be all it would take to win this job.

You are in a difficult position. If you tell the customer that the job requires more than a simple mechanical tie-in, you could lose the job. It's possible that you are being too cautious, but it's better to be safe than sorry. If you throw in with the other two contractors and bid the job against your better judgment, you could be setting yourself up for a lawsuit when the systems don't perform properly. What are you going to do?

In this position, I would suggest that the homeowner either allow me to return with one of my experts or that some expert be contacted by the homeowner directly. I would explain my concerns and the repercussions of putting too many demands on an undersized system. If the homeowner chose to argue with me after this type of educational explanation, I would decline any interest in bidding the job.

Does this story sound a little farfetched? Well, it's not. I've run into circumstances similar to those in the story. My background gives me a broader knowledge of mechanical equipment than what most remodelers have. This advantage has proved useful over the years. There have been several times when customers have told me that contractors had assured them that existing ducts for their heating systems could be used to convey cool air from a new air conditioner. This is rarely the case. Oh, the air will find its way through the ducts, but cooling will not be efficient. Duct sizes are larger for air conditioners than they are for furnaces. A forced-air furnace could be tied into ducts serving an air conditioner, but the reverse is rarely true (Figures 16.1 through 16.4).

The size of duct work is not the only potential problem that remodelers run into. There are many times when an existing heating or cooling system will be adequate to tap into, but there are also plenty of times when they won't be.

How can you tell if an existing system is suitable for expanded use? The best way is to have a couple of experts check the system out. They will have to do a heat-gain, heat-loss worksheet to determine accurately what size system will be needed. Experienced heating-and-cooling mechanics can often make very educated guesses, but be careful not to accept these guesstimates as gospel. I know that many good contractors can eyeball a system and tell if it is capable of taking on extra duty. However, if you don't want to wind up in a bind, get the experts to do a full-blown worksheet on the job and ask for

CFM*	Round duct size	Rectangular duct size
250	9 inch	6" × 10" 8" × 8" 4" × 16"
275	9 inch	4" × 20" 8" × 8" 7" × 10" 5" × 15" 6" × 12"
300	10 inch	6" × 14" 8" × 10" 7" × 12"
350	10 inch	5" × 20" 6" × 16" 9" × 10"
400	12 inch	6" × 18" 10" × 10" 9" × 12"
450	12 inch	6" × 20" 8" × 14" 9" × 12" 10" × 11"

*CFM = cubic feet per minute

FIGURE 16.1 Conversion table for residential duct-work (for 9-inch to 12-inch round brand ducts).

CFM*	Round duct size	Rectangular duct size
400	10 inch	4" × 20" 7" × 10" 6" × 12" 8" × 9"
450	10 inch	5" × 20" 6" × 16" 9" × 10" 8" × 12"
500	10 inch	10" × 10" 6" × 18" 8" × 12" 7" × 14"
600	12 inch	6" × 20" 7" × 18" 8" × 16" 10" × 12"
800	12 inch	8" × 18" 9" × 15" 10" × 14" 12" × 12"
1000	14 inch	10" × 18" 12" × 14" 8" × 24"

*CFM = cubic feet per minute

FIGURE 16.2 Conversion table for residential ductwork (for 10-inch to 14-inch round main or trunk ducts).

their recommendations in writing. This type of action will help keep you off the hook if things don't work out just the way you would like for them to.

There is no rule-of-thumb method used for sizing heating and cooling systems. Some contractors size them with tight limits. This is especially true in track housing, where every dime counts. A lot of contractors, like myself, install systems that are a little larger than they need to be. This provides a margin of error to compensate for any miscalculations. These oversized systems can often handle some extra load, but don't expect any existing system to be substantial enough to take on a large addition or an attic conversion. It is possible that an existing system can manage these types of improvements, but the occasions will be rare.

I've seen contractors push heating and cooling systems to the max. This isn't a good idea. If you overload a system, there is going to come a time when the system doesn't work to its expected performance. When this happens, some angry homeowner is going to be calling you with complaints. This problem can be avoided by simply doing the job right the first time. If you have to go back, at your own expense, to do a retrofit, your profit will be out the window.

It is common for homeowners to put remodelers in tough spots. How many times have you gone out to look at a job and had the homeowners ask you for a guess on what the estimate will amount to? Is it feasible for a contractor to give someone an off-the-cuff price for a basement conversion, an attic conversion, a complete kitchen remodel, or any other type of major work? No, it isn't. Yet, a lot of contractors do it. Why do they feel compelled to make rash comments? I don't know, but I've seen a lot of contractors do it.

There is a big difference between making a rough estimate on the spot and having a preplanned price in mind. Some types of home improvements can be sold right on the spot. Decks, for example, can be figured on a square-footage

CFM*	Round duct size	Rectangular duct size
50	4 inch	4" × 4"
75	5 inch	4" × 5"
		4" × 6"
100	6 inch	4" × 8"
		5" × 6"
125	6 inch	4" × 8"
		5" × 6"
		6" × 6"
150	7 inch	4" × 10"
		5" × 8"
		6" × 6"
175	7 inch	5" × 10"
		6" × 8"
		4" × 14"
		7" × 7"
200	8 inch	5" × 10"
		6" × 8"
		4" × 14"
		7" × 7"
225	8 inch	5" × 12"
		7" × 8"
		6" × 10"

*CFM = cubic feet per minute

FIGURE 16.3 Conversion table for residential ductwork (for 4-inch to 8-inch round brand ducts).

CFM*	Round duct size	Rectangular duct size
1200	16 inch	10" × 20"
		12" × 18"
		14" × 15"
1400	16 inch	10" × 25"
		12" × 20"
		14" × 18"
		15" × 16"
1600	18 inch	10" × 30"
		15" × 18"
		14" × 20"
1800	20 inch	10" × 35"
		15" × 20"
		16" × 19"
		12" × 30"
		14" × 25"
2000	20 inch	10" × 40"
		12" × 30"
		15" × 25"
		18" × 20"

*CFM = cubic feet per minute

FIGURE 16.4 Conversion table for residential ductwork (for 16-inch to 20-inch round main or trunk ducts).

basis with enough accuracy to make selling one right on the spot feasible. Room additions can even be figured out in advance, but you have no way of knowing what to expect from a heating or cooling system until you see it. If your company uses a per-square-foot price for room additions, it must be based on being able to tie into existing mechanical systems.

If you quote a price for an addition and then find that the existing mechanical systems won't handle it, who do you think is going to be forced into paying for a new system? Probably your company.

Duct Work

Duct work will be one of the first considerations when evaluating an air-conditioning system or a forced hot-air heating system. Someone is going to have to determine if the existing ducts can be tied into effectively to serve new living space. This someone should be an expert, and preferably the one who is giving you a firm quote for doing the work. Don't attempt to evaluate duct work on your own, unless you have a much higher level of knowledge about such systems than an average remodeling contractor.

There are some basics about duct work that you can look for. When duct work leaves a plenum (this is the duct work located in the immediate area of the furnace), it will normally leave as either a trunk line or as individual ducts. If the ducts coming off the plenum are small, you are looking at individual supply ducts. More likely, you will see a large, rectangular duct extending for some distance. Smaller ducts will take off from this main trunk. As the trunk line becomes longer, it should also become smaller. To maintain a proper air flow, the size of a trunk line has to be reduced as it becomes longer in length. If you see a trunk line that is not reduced as it runs most of the length of a home, you can expect to have air-flow problems. It would seem that this problem would be rare, but it is not all that uncommon.

I recently rented a house while building my new home. The house I rented suffered from an oversized trunk line. Whoever installed the duct work did a poor job. Because the trunk line was not reduced progressively, rooms at the far end of the main duct were never heated as well as rooms closer to the origination point of the trunk line. Because the duct was too big, there was no opportunity for a volume and pressure of air to exhaust through the heat registers. This resulted in cold rooms. To get the cold rooms warm, the thermostat had to be set so high that other rooms in the house were too hot.

If you came into a house, like the one I had rented, and added to the existing trunk line, your new installation would not perform well. The customer might accept its performance, because some of the other parts of the home would probably be affected similarly, but you shouldn't set yourself up for the risk. If you see a major trunk line that is running full size for a long length, call in an expert to evaluate the needs for making the job right, and discuss the problem with your customer, before any work is done.

The main trunk line is the only portion of duct work that should have an affect on any new work that you do. However, if the seams on the trunk line are not sealed properly, air can escape. This reduces the air flow and the effectiveness of the trunk line. Ultimately, it can affect your new tie-ins. They will not receive the amount of air that they should. If you see gaps at seams in the trunk line, discuss the problem with your customer before you accept the job, and detail a release of liability for the potential problems associated with existing conditions.

We've talked previously about how duct work for an air-conditioning system is usually larger than that used for a heating system. This is, perhaps, one of the most likely traps for a remodeler to fall into. If a customer wants you to tie a new air-conditioning system into existing ducts, have the size of the ducts evaluated by an expert. Taking a job on face value, without accurate sizing data to go by, can result in major problems for you.

The Main Unit

The main unit of a heating or air-conditioning system is an expensive component. While you should not be responsible for the overall condition of this unit, you could be held accountable for not knowing or notifying the customer that the existing unit is inadequate for the additional load you are creating for

- Have a load calculation done for cooling and heating needs.
- Shop for a desirable brand and type of heat pump.
- Decide on air terminal requirements.
- Evaluate outdoor air requirements.
- Choose duct locations and have ductwork sized.
- Choose locations for indoor and outdoor equipment.
- Confirm that installation requirements can be met.
- Choose equipment controls.
- Evaluate initial cost of the system.
- Evaluate efficiency and operating costs for the system.

FIGURE 16.5   Basic steps in choosing a heat pump.

it. This is, again, a time to call in an expert. Whether the main unit is a boiler, a furnace, or a heat pump (Figure 16.5), you should have a qualified professional evaluate it. Even if you are making only modest increases in living space, you could be setting yourself up for big trouble if you don't document that the existing unit can handle the new load.

Radiators

Depending upon where you work, radiators might still be very much in use. Older houses frequently depend on radiators for their heat. If you are estimating a job where the customer wants radiators installed by you, be careful. Radiators are sometimes hard to come by, and they are never cheap. Before you make any commitment for installing radiators, confirm their price and availability.

Electric Heat

Electric heat is simple enough to understand. It's so simple, in fact, that some remodelers find themselves in trouble because of it. Let's say that you are doing an estimate for a basement conversion. All of the upper level of the home is heated with electric baseboard heat. You've discussed heating options with the customer, and they are willing to stick with electric heat for the new living space. Your job seems simple. Hang a few baseboard units, run a little wire, and bingo, the heat work is done. This might be true, but don't count on it. Suppose the main electrical service is full and will not accept any new circuits? Who is going to pay for adding a new set-up for the additional heat? If you take the job before discovering that the box is filled to capacity, you might very well be eating the cost of an additional service. Check the panel box to make sure there is room to grow before you commit to installing new electric heat.

OPTIONS

What are your options for heating or cooling new living space? Is planning for an addition the same as planning for an attic conversion? Does a basement conversion offer challenges that other types of living space doesn't? How are you going to heat and cool the room you are building over that garage?

Designing and planning heating and cooling systems for new living space is a job that should be left to experts. If you are a general remodeling contractor, you are not likely to possess the skills needed to create a near-perfect design. However, you are the person who most homeowners will talk to about their options, and this means you must prepare yourself for their questions.

It is best to leave detailed technical questions for the experts, but you should be able to make some general recommendations on your own. To prepare you for this, let's look at some common questions you might be asked.

My Basement

I want to finish off my basement, can you tell me what is the best way to heat it? This is a loaded question. Finding a best method for heating a basement might depend heavily on the customer's personal opinion of what "best" is. Is cheap installation cost best, or is inexpensive operating cost best? Either could be, depending upon the circumstances.

For example, if someone is finishing off their basement as a game room that will be used only on weekends, the heating requirements during the week are minimal. It could make a lot of sense to install electric baseboard heat for this room, because electric heat is the cheapest of all types to install. Electric heat can be very expensive to operate; however, if it is only be called upon once of twice a week, the high operation cost, on a per-hour basis, may not create a problem. Before you can tell a customer what's best, you have to know what the customer is trying to achieve. This means that, for every question you are asked, you have to ask some yourself.

Basement conversions offer a few challenges that other types of rooms don't normally present. For one thing, the floor of a basement makes it extremely difficult to route heating pipes or ducts through or under it. Some basements have low ceilings to begin with, dropping the ceiling height farther, to accommodate duct work, can present a real problem. If the exterior walls of a basement are furred out with thin material, putting heating ducts or pipes in them can be nearly impossible. Plus, the cold produced around a buried masonry wall can have an adverse effect on heating equipment.

One advantage to a basement is that the main heating unit for the whole house will probably be located in the basement. This makes access to the unit easier than it would be if an attic conversion was being done. It might very well be possible to run individual ducts through joist bays and down interior partition walls. Because there is often a girder somewhere near the center of a basement, it might be that a trunk line is already chasing along the beam. Because the beam will probably be boxed in during the conversion process, it would make sense to run duct work along the beam and enclose it in the same box.

Electric heat is an inexpensive option to install. Assuming that the electrical box is adequate for new baseboard heating units, electric heat is simple and cost-effective to install. However, paying for the operation costs of running electric heat can get very expensive. The long-term effect of electric heat might prove that its low initial cost is overshadowed by its high cost of operation.

When you are sitting down with customers in an attempt to work through plans and design issues, it is important to look at all angles. It is unreasonable to assume that any one person is an authority on all subjects. This certainly applies to a remodeling contractor. If you are a carpenter by trade, it is unfair to expect your knowledge of heating and air conditioning to be on par with that of a person who has developed in the HVAC trade. This doesn't mean that you shouldn't have a cursory understanding of the trade. You should. It will pay dividends throughout your career to know as much about all the of trades as you can.

How Will You Get Them up There?

I need to have new heating and air-conditioning ducts run into my attic, how will you get them up there? This question often arises around attic conversions. Duct work is cumbersome in its size. Unlike small plumbing pipes and electrical wire, ducts can be difficult to route through a finished home.

The most common method for getting new ducts into an attic is the use of chases. Sometimes the chase is built in the corner of a room. They are often built inside of closets. It is common for chases to be used to disguise the location of duct work.

I Want a New Family Room.

I want a new family room built onto my house, but I don't think my existing heat pump is large enough to handle it. This is a statement that you might run into. If the existing heat pump is too small to heat and cool the addition, you could offer to install a one-piece heat pump that would serve only the addition. This would be less expensive than replacing the existing heat pump with a larger one.

I Have Forced Hot-air Heat.

I have forced hot-air heat in my home, but I don't like the dust associated with it. This statement is not uncommon among homeowners. Forced-air systems do displace a lot of dust. If you are in the middle of a major remodeling job, the homeowner might be considering a replacement heating system. What would you recommend?

If a customer is dead-set against any type of duct work and forced-air system, you can offer a hot-water system. Hot-water baseboard heat is very popular in areas where winter temperatures are extremely cold. These systems are not inexpensive, but they are very good at combating cold, and they don't blow dust all around. Electric heat is another option to consider, but its operating cost might be prohibitive, depending upon the geographical location of the home. A house is Florida might do very well with electric heat, where a house in Maine would cost a small fortune to heat with electricity.

HEAT PUMPS

Heat pumps are, in large part, the most popular type of HVAC used today. They provide both heating and cooling at an affordable cost. For this reason, I am providing a wealth of tables and charts pertaining to heat pumps. This information should help you in your contracting decisions and work. (See Figures 16.6 through 16.37)

- Air-to-air
- Air-to-water
- Water-to-water
- Water-to-air

FIGURE 16.6 Common types of heat pumps.

- Solar-powered heat pumps
- Passive solar heat pumps
- Solar-assisted heat pumps
- Solar evaporator coils

FIGURE 16.7 Types of solar-powered heat-pump systems.

Characteristic	Rating
Suitability	Good
Stability	Extreme
Availability	Excellent
Initial cost	Low
Operating cost	High
Drawbacks	Frosting

FIGURE 16.8 Profile of an air-source heat pump.

Characteristic	Rating
Suitability	Good
Stability	Extreme
Availability	Excellent
Initial cost	Mid-range to high
Operating cost	Low
Drawbacks	Complicated system with expensive set-up costs

FIGURE 16.9   Profile of a solar-source heat pump.

Characteristic	Rating
Suitability	Varies
Stability	Fair
Availability	Limited
Initial cost	Mid-range
Operating cost	Low
Drawbacks	Corrosion and dry spells

FIGURE 16.10   Profile of a surface-water source heat pump.

Characteristic	Rating
Suitability	Excellent
Stability	Stable
Availability	Very good
Initial cost	Mid-range
Operating cost	Low
Drawbacks	Mineral build-ups

FIGURE 16.11   Profile of a well-water source heat pump.

Characteristic	Rating
Suitability	Good
Stability	Stable
Availability	Excellent
Initial cost	Mid-range
Operating cost	Low to medium
Drawbacks	Leaks are hard to find and expensive to repair

FIGURE 16.12 Profile of an earth-source heat pump.

- Low operating costs
- Provides a high COP from the heat pump
- When coupled with storage facilities, it can operate at off-peak electric rates

FIGURE 16.13 Advantages of solar-assisted heat pumps.

- No moving parts to break or wear out
- Low cost to operate
- Few defrost cycles
- Increases heat pump COP
- No wind chill factor
- Minimum auxiliary heat required

FIGURE 16.14 Advantages of passive solar systems.

- Provides a high source of evaporator heat
- Provides a high COP from the heat pump
- Minimum auxiliary heat required

FIGURE 16.15 Advantages of solar evaporator coils.

- No heating costs
- Simplicity in design
- Low maintenance costs
- Longevity

FIGURE 16.16 Advantages of water-source heat pumps.

- Potential for well or surface water drying up
- Mineral build-ups
- Corrosion
- Possible water pump failure

FIGURE 16.17 Disadvantages of water-source heat pumps.

- High initial costs
- Complex design
- Not efficient on cloudy days
- Potential for high maintenance costs

FIGURE 16.18 Disadvantages of solar-assisted heat pumps.

- Space requirements can present problems
- No other prominent disadvantages

FIGURE 16.19 Disadvantages of passive solar systems.

- Requires sunlight for best results
- Minimum auxiliary heat required

FIGURE 16.20 Disadvantages of solar evaporator coils.

Problem: Heat pump will not run.
- Check fuses and circuit breakers.
- Check for broken or loose electrical wires.
- Check for a possible low voltage supply or circuit.
- Check thermostat.

FIGURE 16.21 Quick check troubleshooting for water-source heat pumps.

Problem: Heat pump is short cycling.

- Check for a compressor overload.
- Check electrical wiring.
- Check thermostat for location and defects.
- Check lockout relay.
- Check high-pressure cut-out.
- Check discharge pressure to see if it is too high.
- Check refrigerant charge.
- Check for inadequate water flow.
- Check for excessive air flow.
- Check for defective high-pressure switch.

FIGURE 16.22 Quick check troubleshooting for water-source heat pumps.

Problem: Heat pump will produce only heat.

- Check for a defective reversing valve.

FIGURE 16.23 Quick check troubleshooting for water-source heat pumps.

Problem: Heat pump will not run when calling for heat.

- Check for a clogged air filter.
- Check for an improperly set thermostat.
- Check for a defective thermostat.
- Check for a defective blower motor.
- Check for problems with the electrical wiring.

FIGURE 16.24 Quick check troubleshooting for water-source heat pumps.

Problem: Heat pump produces inadequate heating or cooling.

- Check for leaks.
- Check to be sure the heat pump is not too small.
- Check to see that sufficient water pressure is available.
- Check to see that thermostat is not improperly located.
- Check for inadequate air flow.
- Check for a lack of refrigerant.
- Check for a defective compressor.
- Confirm that the blower is not blowing in reverse.
- Check for a defective reversing valve.
- Check operating pressure.
- Check refrigerant system.

FIGURE 16.25 Quick check troubleshooting for water-source heat pumps.

Problem: Heat pump is off on low-pressure cut-out control.

- Check refrigerant charge.
- Check for low suction pressure on cooling cycle.
- Check for low suction pressure on heating cycle.
- Check for defective low-pressure switch.

FIGURE 16.26   Quick check troubleshooting for water-source heat pumps.

Problem: Heat pump is not cooling properly and is indicating a high suction pressure.

- Check for defective compressor valves.
- Check for defective reversing valve.
- Check refrigerant charge to see if it is too high.
- Check for noncondensibles in the system.

FIGURE 16.27   Quick check troubleshooting for air-source heat pumps.

Problem: Heat pump provides heat when it should be cooling.

- Check for a defective reversing valve.
- Check for defective electric resistance elements.

FIGURE 16.28   Quick check troubleshooting for air-source heat pumps.

Problem: Heat pump is not heating properly and is indicating a high liquid pressure.

- Check for restricted air flow.
- Check refrigerant charge to see if it is too high.
- Check for dirty filters.
- Check for a dirty coil.
- Check the blower motor.
- Check for noncondensibles in the system.

FIGURE 16.29   Quick check troubleshooting for air-source heat pumps.

Problem: Heat pump will not defrost.

- Check for closed relays on the reversing valve.
- Check for closed defrost relays.
- Check for obstructions in the pressure tube.

FIGURE 16.30 Quick check troubleshooting for air-source heat pumps.

Problem: Heat pump is not cooling properly and its operating pressures are not normal.

- Check for problems in the electrical system.
- Confirm that the heat pump is of a proper size.
- Check for air leaks.

FIGURE 16.31 Quick check troubleshooting for air-source heat pumps.

Problem: Heat pump is not heating properly and is indicating a low suction pressure.

- Check for a defective expansion valve.
- Check for a defective outdoor fan or motor.
- Check refrigerant charge.
- Check for restricted tubes at outdoor coil.
- Check for a dirty outdoor coil.

FIGURE 16.32 Quick check troubleshooting for air-source heat pumps.

Problem: Defrost cycle will not stop.

- Check for defective defrost controls.
- Check for a defective temperature bulb.
- Check the refrigerant charge.
- Check for a dirty indoor coil.
- Check for dirty air filters.
- Check for a defective expansion valve.

FIGURE 16.33 Quick check troubleshooting for air-source heat pumps.

Problem: Heat pump will not run.

- Check fuses and circuit breakers.
- Check for faulty wiring.
- Check thermostat setting, location, and calibration.
- Check compressor overloads to see if contacts are open.
- Check to see if high-pressure control is open.
- Check for defective transformer.

FIGURE 16.34 Quick check troubleshooting for air-source heat pumps.

Problem: Heat pump is not cooling properly and is indicating a high liquid pressure.

- Check for a dirty outdoor coil.
- Check to see that outdoor fan is operating properly.
- Check refrigerant charge to see if it is too high.
- Check for noncondensables in the system.
- Check for restrictions in the liquid line.

FIGURE 16.35 Quick check troubleshooting for air-source heat pumps.

Problem: Heat pump is not heating properly and its operating pressures are not normal.

- Check outdoor thermometer for improper settings.
- Check for restricted air flow.

FIGURE 16.36 Quick check troubleshooting for air-source heat pumps.

Problem: Heat pump will not defrost completely.

- Check refrigerant charge.
- Check defrost circuitry.
- Check defrost control.
- Check compressor valves.

FIGURE 16.37 Quick check troubleshooting for air-source heat pumps.

RELY ON LICENSED PROFESSIONALS

It is always best to rely on licensed professionals when dealing with HVAC issues. These contractors should know their trade better than you do. However, calling in a mechanical contractor is not the end of your responsibility. As the general contractor, it will be up to you to coordinate the planning and installation of heating and air-conditioning systems. For example, if a chase is going to be needed to hide duct work, you have to know about it and plan for it.

It is essential that you have good communication and a close working relationship with all of your subcontractors. When one trade is working outside of the communication circle, plenty can go wrong.

Call in experts for your HVAC needs, but stay informed and keep a watchful eye on the job. Remember, the customer is going to hold you responsible for the work done by your subcontractors.

CHAPTER 17
PLUMBING SYSTEMS

Plumbing is one area of work where a lot of remodelers and some builders lose money. Existing conditions apply to almost all types of work, but they might apply more to plumbing and carpentry than any other types. Many remodeling contractors begin their careers as carpenters. As they learn their trade, they venture out on their own, taking with them a good working knowledge of carpentry.

While being prepared to deal with a multitude of potential problems in their established fields, these contractors are far from prepared to deal with plumbing systems. They've seen them, and they know something about them, but they don't know enough to keep themselves out of trouble. This can also be true of electrical wiring and other existing conditions. However, plumbing seems to be the number-one downfall for most remodelers who are bidding jobs without the assistance of individual experts. Part of this problem might be that plumbing can seem simple, when in reality, the codes pertaining to plumbing are quite complex.

I've seen jobs where remodelers grossly underestimated the costs involved for new plumbing. Many of these situations could be identified as being a lack of knowledge on the remodeler's part. If these contractors had called in a competent plumber before signing a contract with a homeowner, most of the losses could have been avoided. However, for some reason, many contractors wait until they have been awarded a job to call in a plumber. This is true, too, of building contractors. By this time, if there is a problem, the contractor must absorb the financial loss. Whenever plumbing is involved in a remodeling job or new construction, the price of the work can go way up, so don't allow yourself to get caught on a bad financial updraft.

How can you avoid problems with plumbing? You can't always avoid the problems, but you can shelter yourself from them. You can do this by having experts look over existing conditions and future plans. Experienced plumbers can look at a set of blueprints for new construction and point out numerous potential problems. You can further protect yourself with liability clauses in your contracts. Another way to reduce your risk is to learn everything you can about plumbing and the codes that govern its installation. This is advisable even if you do rely on outside professionals for all your plumbing needs.

Because plumbing is a licensed trade, not just anyone can install plumbing, legally. This means that you will most likely be calling in plumbing contractors for your jobs. Some large remodeling firms have their own inhouse plumbers on payroll and so do some developers and builders, but this option is too expensive for most contractors. Subcontractors are typically the most cost-effective way to get your plumbing done.

When you go out to price up a job, do you take your entire stable of subcontractors with you? Most contractors don't. Taking a representative of each trade with you on estimate calls is impractical. It's a good idea, but one that doesn't work well in the real world. A better approach is to make the first estimate call on your own, and then bring in the specialized trades to look over areas of the job where you are not comfortable with your own assessment. This should, however, be done before any estimate or quote is given to the customer.

THE PLUMBING CODE

The plumbing code is a complicated thing to understand. Many professional plumbers don't understand it, and some only understand portions of it that apply most to what they do. So, if plumbers have trouble with the code, how can you expect to understand it? Well, the code does consist of some cryptic language, but most of it can be broken down into readable text.

I learned the plumbing code by reading the book on my lunch break. When I had questions about the code, I asked people who would know the answers to explain them to me. In doing this, I learned the entire code in a relatively short period of time. You might find that reading a code book in your off hours is no fun, but it can make you a more profitable contractor.

I have always felt vulnerable when hiring someone to do work for me that I know little about. As a contractor, it used to bother me to hire electricians and other subcontractors who worked in trades that I was not familiar with. Oh, I'd seen plenty of electrical work being done, but I didn't know why it was being done the way it was. The same was true of HVAC work and other trades. If I hired a contractor to paint a house, I wanted to know about how long the job should take and what types of materials should be used. This was the only way I could think of to keep my subs honest.

In feeling this way, I worked hard to learn a lot about all of the trades. I know some of them better than others, but I know all of them well enough to know when a subcontractor is trying to put something over on me. I strongly recommend that you follow my lead in this matter. Surviving for years as a contractor is difficult enough, but doing it when you don't know what you're doing is much harder. You owe it to yourself to get acquainted with all of the trades.

Plumbing is a good trade to start your new learning cycle with. I'm not saying that you should strive to get a license to install plumbing, but you should learn the basics. If you don't know the difference between a gate valve and a stop-and-waste valve, you could make a mistake in figuring the cost of installing a water heater. A much bigger problem could arise if the house you are about to add a bathroom in has a sewer that is rated for only two toilets, both of which are already installed. Digging up and replacing a sewer is not a cheap proposition. There are a number of potential disasters waiting for you in the plumbing phase of your work as a contractor. Let's talk about some of them.

IN THE KITCHENS

A lot of remodeling goes on in the kitchens of homes. Kitchens are traditionally one of the most profitable rooms for a homeowner to remodel, so they get a lot of attention. There are magazines available where the primary topic throughout the entire publication is kitchen remodeling. With this public display, kitchen remodeling is a booming business. If you are not already doing a lot of this type of work, you might find yourself doing it soon.

When is a kitchen remodeled that some plumbing is not altered? Not very often. Nearly all kitchen jobs involve plumbing. Whenever plumbing is involved, problems might be looming in the background.

Let's put ourselves on a hypothetical job site. We are there to estimate a complete kitchen remodel. The job calls for the existing sink location to be moved 4 feet down the exterior wall. A new island cabinet will be installed, and it will contain a sink for washing vegetables. This island sink is in addition to the standard, double-bowl sink in the main counter. A garbage disposer is wanted, and so is an ice-maker connection. The area under the kitchen is a crawl space, so access is reasonably good. This is a two-story home, and there is a bedroom over the kitchen.

The first thing you do is open the doors on the sink base, to inspect the existing plumbing. There are copper water pipes and plastic drainage fittings. You feel comfortable knowing that the materials are modern. In fact, you see no real problem at all. There is access below the floor, the materials are modern, and the work seems innocent enough. What could go wrong? Well, you are about to find out.

To point out the potential problems surrounding this job, let's take the information we have gained and examine it one line at a time. The first order of business is the relocation of the existing sink location. Did you check to see that the existing sink was vented properly? No, you didn't. When you relocate a fixture, a permit is required. It is also standard procedure for all major work, such as a fixture relocation, to trigger compliance with current codes. The old sink might not be vented or maybe it was vented with a mechanical vent. While this would not have been a problem if all you were doing was changing sinks, the fact that you are changing the sink location makes all the difference in the world. All of a sudden, you might have to extend a vent from the sink to a point outside the roof of the home. This would obviously require some work that you would not normally plan on in a simple kitchen remodel. However, this is not the only potential problem waiting for you.

The customer wants a new island sink installed. Do you know how to vent an island sink? Are you aware of the pipe size that is required for this type of combination drain and vent work? Did you even know that the sink is required to be vented? If you want to survive as a contractor, you'd better know information like this. Island sinks are complicated to vent. They require access to a vent that either penetrates a roof or that is tied into a vent that does. This might be the existing vent in the kitchen, if there is one. However, the pipe size could still get you. Your plumber might have to go way back under the house to find a drain large enough to tap into for the island sink. This, of course, will require running new, larger piping to the sink location. You probably thought that the new sink could be tied into the pipe under the flow that serves the present sink, but this might not be the case. See how your losses are building? We're not done yet.

Do you remember the request for a garbage disposer? Did you give that much thought? I'm sure you probably figured it would be a simple installation, because the piping was all so modern. Well, the hook-up would be simple, but can the customer have a disposer? Ah, you hadn't thought of this, had you? Some plumbing codes prohibit the installation of a garbage disposer where the drain will empty into a septic tank. This is not a universal rule, but it does apply in many places. If the house you are working up an estimate for is served by a private waste-disposal system, you might not be allowed, by code, to have a garbage disposer installed. If this is the case, you should make the customers aware of it promptly.

There is another catch in the code that could make your cost for installing a disposer more than you expect it to be. Depending upon what your local plumbing code requires, your plumber might have to upgrade the drainage pipe for the disposer to a larger size. Most jurisdictions require a kitchen drain that will serve a disposer to be larger in diameter than what is required just for a sink drain. Again, more money is lost.

Now, what about the icemaker? Well, the icemaker is innocent enough. Your plumber should have no trouble tapping into the cold-water pipe under the house. This will provide a water source for the icemaker, and running the tubing in a crawl space should not present any unusual difficulties.

So, out of four plumbing items, three of them carry potential risk beyond what most remodelers would think of. There is still one more potential problem that we have not discussed.

Plumbing codes regulate the number of fixtures that can be served by various sizes of pipes. Typically, no more than two fixtures can be served with hot and cold water from a ½-inch pipe. When you looked under the sink to check out the piping, you noticed that the pipes were copper, and we will assume they were ½-inch pipes, because they are so common under kitchen sinks. What you didn't do is trace the pipes to see if other fixtures were being served by them. If any other existing fixture is getting its water from these pipes, installing the island sink will require upgrading the supply pipes to ¾-inch material. Can you see how some simple plumbing can turn your profit picture into a nightmare?

ADDING A NEW BATHROOM

Adding a new bathroom to a home can become very involved and quite expensive. The job might entail substantially more work than what one would first assume. In the following examples, we will talk only about issues pertaining to plumbing. There is, of course, a lot more to adding a bathroom than just plumbing, but we should stay focused on the issue at hand.

A Basement Bath

What's involved with installing a basement bath? A lot of people want one, and there's good money to be made by installing them. However, there are some red flags that should go up when you begin your estimating process. If you already know what these indicators are, you are ahead of the pack. For those who don't know what they are getting into with a basement bath, let's look at an example.

Whenever a basement bath is added, the concrete floor must be broken up, unless the bathroom was roughed-in during the home's construction. This should come as no surprise to you. However, what are the details involved with the drainage and vent system for a basement bath?

Let's assume that you have been called out to figure a job where a full-bath is wanted in the basement. You inspect the job and see that the sewer for the home leaves the basement at a height of about 3 feet above the finished floor level. This means there are no drains in the floor that are suitable for connecting the bathroom to. You now know that a pump system will be required. This evaluation should be easy enough for any experienced remodeler to read. However, what is involved with a pump system?

The floor will have to be broken up to accept a sump basin. The size of the basin will vary, but it will usually be about 30 inches deep and about 18 inches in diameter. Once the basin is in place, drains will run under the concrete floor to the various fixture locations. This will require cutting the concrete and trenching the ground. This is pretty simple with the use of a jackhammer, unless you find something you didn't expect. This is a point that you should cover in your contract. I got burned on a job once because I didn't address underground obstacles.

Years ago, we were installing a basement bath in a home in Virginia. When we broke the floor up and started to dig through the crushed stone, we found water. I'm not talking about a little puddle of ground water, we hit an underground stream. The water moved so rapidly that it was washing away the stones that we had disturbed. Pumping the water was out of the question. It was coming in faster than we could pump it. Our plumbing trenches were flooded. As you can imagine, glue joints on plastic pipe don't set up well when they are submerged in water. With no other option, we had to pipe the drains with cast-iron pipe. This allowed us to get the job done, but the cost for doing this work was much more, both in labor and material, than what it would have been if the job had been done with plastic, as I had planned. These little unforeseen adventures can erode your profit quickly.

Assuming that existing conditions don't throw you any curves, a basement bath is pretty straightforward. As long as you realize up front that a pump station is needed, you've got a good head start. However, suppose the sewer is too small? Most plumbing codes will not allow more than two toilets to be installed on a single 3-inch drain. Many tract houses are plumbing with 3-inch sewers, to keep the construction cost down. A good number of the houses already have two toilets, so no others can be added to the sewer. If you come into one of these homes and agree to install a full-bath, you could be in a world of trouble. Before you give anyone a price for installing a new bathroom, check the size of their sewer and the local code requirements. If the sewer has a 4-inch diameter, you're safe, but beware of 3-inch sewers.

Upstairs and Attic Bathrooms

Upstairs and attic bathrooms don't require pump systems, but they can still be affected by the size of the building sewer. Always check the diameter of the building sewer before you commit to installing a new toilet. The biggest problem with an attic bath is usually hinged on getting pipes up to the attic. Some plumbers, in some houses, can do this without any destruction to existing wall coverings. Most jobs, however, will require either the opening of walls or the building of chases for the pipes to be installed. This is something that you should be aware of from a budget point of view, and it is an issue you should cover with the customer before someone walks into their home and sees a plumber hacking away at their fine wallpaper.

Another problem that you are likely to encounter with an attic bath is the floor level. Plumbing pipes have to be graded for adequate drainage. This grade is usually set at ¼ inch per foot of pipe. In other words, if you have a pipe running 20 feet in length, one end of the pipe will be 5 inches higher than the other end. This can give you some trouble in an attic, where headroom is at a premium to begin with. Make sure that you and your plumber plan the installation to avoid running out of room for the pipe grade.

There is one other thing along this line that I should mention. When you are figuring the rise of a pipe for a toilet, don't forget to add height for the turn-up and the closet flange. A 3-inch elbow and a street flange can consume a lot of space, and this height is in addition to the drain pipe.

Slipping in a Half-bath

Slipping in a half-bath somewhere in a house is a common remodeling project. The powder room might be placed under a set of stairs, or a closet might be converted to accommodate it. Either way, a 2-inch vent is going to be needed for the plumbing. This vent must either rise to a level 6 inches above the flood-level rim and tie into another vent or it must extend through the roof of the home. This can be a big deal if you haven't allowed for the extra work in your cost projections.

ADDITIONS

Additions sometimes contain plumbing. The installation might be a wet bar or a full-bath, but either way, the plumbing will need a water supply and a drainage outlet. This can be quite difficult in some cases.

Water pipes are more flexible in their installation than drains are, so most of your big problems will be with the drains. Some additions are built in a way that makes tying into an existing drainage system nearly impossible. For these jobs, plumbers often run the drain from the addition underground until it intercepts with the sewer for the property. This can be effective, but it also tears up a customer's lawn. If you anticipate doing this, you had better prepare the customer for the trauma. You had also better figure in the cost of excavating equipment and landscaping repairs, because these items can amount to substantial money.

Not all additions require extensive outside plumbing work. Some of them are built so that the drains can run under the addition, penetrate the home's existing foundation wall, and tie into the existing building drain. This is something that you will have to evaluate on each job. If outside work is required, you can't afford to overlook the cost.

OLD PLUMBING

Old plumbing can present a lot of problems for remodelers. For example, old galvanized steel pipe tends to clog up with rust and other deposits. This happens whether it has been used to convey potable water or drainage. Anytime you authorize your plumber to tie into existing galvanized pipes, you could be setting yourself up for call-backs.

My experience has shown that all galvanized pipe should be replaced with modern plumbing materials. The up-front savings of tying into old steel pipes is generally lost in unbillable time for call-backs. This doesn't even take into account the frustration of homeowners who have just paid handsomely for quality remodeling work that doesn't function properly. My advice is this, if you see galvanized pipe on a job, talk the customer into replacing it.

To illustrate how you can become responsible for existing piping problems, let me share some of my past experience. I've done a lot of kitchen remodeling. Many of the jobs have entailed the addition of garbage disposers. In my earlier years, I used to work right from the trap arm (the pipe that sticks out under the kitchen sink). If the trap arm was galvanized, I would have my people connect to it. It didn't take more than a couple of jobs for me to change my philosophy on this issue.

I discovered that old galvanized drains that would adequately drain a sink-full of water would not necessarily take on the discharge of a garbage disposer. This experience came at a high price. The sinks would test out okay when they were installed, because only water was going down the drain. However, as customers used their disposers, the drains would begin to clog up. Snaking the drains would punch holes in the clogs and get the drainage running again, but the repairs didn't last long.

Invariably, the galvanized pipe had to be cut out and replaced with plastic pipe. This, of course, frequently involved cutting into walls and messing up brand-new remodeling jobs. Customers don't appreciate this type of action. Anyway, I learned quickly to have all galvanized drains replaced, even if it had to be done at my own expense. It was cheaper to replace them during the remodeling work than it was after, and the customers never got mad under these conditions.

Some other types of old drainage pipes that you are likely to run across are DWV copper and cast iron. Both of these materials make good drains, and they last for a very long time. Unless circumstances are unusual, there will be no need to replace either of these types of drains.

Lead pipe is not found in a lot of homes today, but it still turns up from time to time. The most common locations are near the traps of bathtubs, and the bends under toilets. If you notice any lead pipe, plan on replacing it. This material is soft and does not adapt well to remodeling work. Vibrations from cutting out floors, building new walls, and so forth are likely to make the old lead joints leak. Even the lead itself is likely to crack and leak. Count on re-placing any lead you find.

Brass pipe was used to convey potable water in the years past. This pipe is not all bad, but it is not as desirable as modern materials. I won't say that you have to replace all brass pipe that you find, but if you have to tap into a brass system, you should be prepared for some problems. The screw joints along the system might develop leaks as you monkey around with fitting in new tees. If I saw brass pipe on a job, I would make sure that the customer was not going to hold me responsible for stress leaks down the line.

Galvanized steel pipe that is used for water pipe is just about as bad as the same pipe used for drains. I've seen galvanized water lines seize up with so much of a collection of debris, such as rust, that water would only trickle out of the end of it. If you plan to ask your plumber to work with old galvanized water piping, be prepared to pay a steep price.

Old cut-off valves can complicate your plumber's life. If the main cut-off valve in a house won't stop the flow of water, your plumber might be forced to cut the water off at the street. This usually isn't a big deal, but it can add up a little in labor. If you know that you will have to tap into existing water lines, you can test the main cut-off while you are doing your estimate inspection. Finding out ahead of time that a valve is defective can save you money and your plumber some time and trouble.

USED MATERIALS

Not many homeowners request contractors to install used materials, but oc-casions arise where this is the case. I've had homeowners who wanted my company to install plumbing fixtures that they had purchased at yard sales and such. In my time in the trades, I've had requests to install just about every type of used residential fixture in remodeling jobs. I don't have a blanket policy against doing this, but I don't like it, and I don't do it very often.

Used fixtures can be trouble just waiting for a place to happen. If you have your plumber install some of them, you might be the lucky one to win the honor of being the place where the trouble surfaces. Old toilets can have hairline cracks that can't be seen easily but that can leak profusely. Sinks offer the same risks. Used water heaters can leak, be full of sediment, have burned-out elements, and have a host of other problems. Bathtubs and showers can also harbor unseen defects. If you want to be safe, refuse to install used plumbing fixtures.

Claw-foot tubs are one type of used fixture that seems to be very popular in some remodeling jobs. A case can be made for installing these units, because they are not common. I urge you, however, to approach any job where used fixtures are requested with care. Write a liability waiver that limits your exposure to the connections made to the fixture. By doing this, if the fixture itself is defective, you and your plumber won't be under the gun.

KEEP YOUR EYES OPEN

When you are estimating a job where plumbing work will be needed, keep your eyes open. If copper pipe is stained green, beware. The pipe might be on the verge of rupturing, due to too much acid in the water. When your plumber tries to cut into the pipe, an entire section might have to be replaced. If the building sewer is too small, you could lose all of your profit installing a new one. Whenever major work is done on a system, most codes require that the entire system be brought into compliance with current codes. This can get extremely expensive. A 40-gallon water heater might be doing just fine under existing conditions; however, if you are building an in-law addition, where additional people will be putting demands on the water heater, the heater might not meet the demands.

You should always involve experts in your estimating process before you give any firm prices to customers. It is possible for you to spot some obvious trouble spots with plumbing systems, but you need a seasoned plumber, one who has worked in remodeling for awhile, to make sure that you stay out of hot water. It might be a little inconvenient to have your plumber tag along with you on estimates, but it can save you a lot of money and embarrassment.

If you are bidding new construction, get a few prices from reputable plumbing companies before you submit a bid. There is a tendency for contractors to assume that plumbing is about the same from one job to the next. This can be the case, but don't count on it. Plumbing is often an expensive and potentially dangerous part of a job, financially, so don't leave yourself out on a limb.

The plumbing code is a complex subject. It is not one that we can delve into, in depth, in this chapter. However, I will provide you with pages upon pages of information that will answer your most-often-asked questions pertaining to the plumbing requirements. Study the following information to make yourself more aware of the plumbing needs in your jobs (Figures 17.1 through 17.53).

Fixture or device	Size (in.)
Bathtub	½
Combination sink and laundry tray	½
Drinking fountain	⅜
Dishwashing machine (domestic)	½
Kitchen sink (domestic)	½
Kitchen sink (commercial)	¾
Lavatory	⅜
Laundry tray (1, 2, or 3 compartments)	½
Shower (single head)	½
Sink (service, slop)	½
Sink (flushing rim)	¾
Urinal (1" flush valve)	1
Urinal (3/4" flush valve)	¾
Urinal (flush tank)	½
Water closet (flush tank)	⅜
Water closet (flush valve)	1
Hose bib	½
Wall hydrant or sill cock	½

FIGURE 17.1 Fixture branch piping: Minimum size.

Type of fixture	Minimum size of vent
Lavatory	1¼"
Drinking fountain	1¼"
Domestic sink	1¼"
Shower stalls (domestic)	1¼"
Bathtub	1¼"
Laundry tray	1¼"
Service sink	1¼"
Water closet	2"
Note: At least one 3" vent must be installed.	

FIGURE 17.2   Vent size.

Size of fixture in inches	Distance trap to vent
1¼	2 ft. 6 in.
1½	3 ft. 6 in.
2	5 ft.
3	6 ft.
4	10 ft.
Note: Figures might vary with local plumbing codes.	

FIGURE 17.3   Fixture trap maximum distances.

Type of fixture	Size (in.)
Clothes washer	2
Bathtub with or without shower	1½
Bidet	1½
Dental unit or cuspidor	1¼
Drinking fountain	1¼
Dishwasher (domestic)	1½
Dishwasher (commercial)	2
Floor drain	2, 3, or 4
Lavatory	1¼
Laundry tray	1½
Shower stall (domestic)	2
Sinks:	
Combination, sink and tray (with disposal unit)	1½
Combination, sink and tray (with one trap)	1½
Domestic with or without disposal unit	1½
Surgeon's	1½
Laboratory	1½
Flushrim or bedpan washer	3
Service	2 or 3
Pot or scullery	2
Soda fountain	1½
Commercial, flat rim, bar, or counter	1½
Wash circular or multiple	1½
Urinals:	
Pedestal	3
Wall-hung	1½ or 2
Trough (per 6-foot section)	1½
Stall	2
Water closet	3

FIGURE 17.4 Common trap sizes.

Diameter of trap arm (inches)	Length from trap to vent
1¼	3' 6"
1½	5'
2	8'
3	10'
4	12'

Note: Might vary with local plumbing codes.

FIGURE 17.5 Trap arm maximum length.

Fixture	Flow rate (gpm)
Ordinary basin faucet	2.0
Self-closing basin faucet	2.5
Sink faucet, ⅜"	4.5
Sink faucet, ½"	4.5
Bathtub faucet	6.0
Laundry tub cock, ½"	5.0
Shower	5.0
Ball cock for water closet	3.0
Flushometer valve for water closet	15–35
Flushometer valve for urinal	15.0
Drinking fountain	.75
Sill cock or wall hydrant	5.0

Note: Figures do not represent the use of water-conservation devices.

FIGURE 17.6 Water flow rate.

Type of outlet	Demand (gpm)
Ordinary lavatory faucet	2.0
Self-closing lavatory faucet	2.5
Sink faucet, ⅜" or ½"	4.5
Sink faucet, ¾"	6.0
Bath faucet, ½"	5.0
Shower head, ½"	5.0
Laundry faucet, ½"	5.0
Ballcock in water closet flush tank	3.0
1" flush valve (25 psi flow pressure)	35.0
1" flush valve (15 psi flow presure)	27.0
¾" flush valve (15 psi flow pressure)	15.0
Drinking fountain jet	0.75
Dishwashing machine (domestic)	4.0
Laundry machine (8 or 16 lb.)	4.0
Aspirator (operating room or laboratory)	2.5
Hose bib or sill cock (½")	5.0

Note: Demands do not take into account the use of water-conservation devices.

FIGURE 17.7 Water demand at individual outlets.

Clean-out plugs will be made of plastic or brass. Brass plugs are to be used only with metallic fittings. Unless they create a hazard, clean-out plugs shall have raised, square heads. If located where a hazard from the raised head might exist, countersunk heads may be used. Zone two requires borosilicate-glass plugs to be used with clean-outs installed on borosilicate pipe.

FIGURE 17.8 Clean-out plugs.

Manufactured pipe nipples are normally made from brass or steel. These nipples range in length from ⅛ of an inch to 12 inches. Nipples must live up to certain standards, but they should be rated and approved before you are able to obtain them.

FIGURE 17.9 Nipples.

A stand-pipe, when installed in zone three, must extend at least 18 inches above its trap, but may not extend more than 30 inches above the trap. Zone two prohibits the stand-pipe from extending more than 4 feet from the trap. Zone one requires the stand-pipe not to exceed a height of more than 2 feet above the trap. Plumbers installing laundry stand-pipes often forget this regulation. When setting your fitting height in the drainage pipe, keep in mind the height limitations on your stand-pipe. Otherwise, your take-off fitting might be too low, or too high, to allow your stand-pipe receptor to be placed at the desired height. Traps for kitchen sinks might not receive the discharge from a laundry tub or clothes washer.

FIGURE 17.10 Stand-pipe height.

Back-water valves are essentially check valves. They are installed in drains and sewers to prevent the backing up of waste and water in the drain or sewer. Back-water valves are required to be readily accessible and installed any time a drainage system is likely to encounter back-ups from the sewer.

The intent behind back-water valves is to prevent sewers from backing up into individual drainage systems. Buildings that have plumbing fixtures below the level of the street, where a main sewer is intalled, are candidates for back-water valves.

FIGURE 17.11 Back-water valves.

- All devices used to treat or convey potable water must be protected against contamination.
- It is not acceptable to install stop-and-waste valves underground.
- If there are two water systems in a building, one potable, one nonpotable, the piping for each system must be marked clearly. The marking can be in the form or a suspended metal tag or a color-code. Zone two requires the pipe to be color-coded and tagged. Nonpotable water piping should not be concealed.
- Hazardous materials, such as chemicals, may not be placed into a potable water system.
- Piping that has been used for a purpose other than conveying potable water may not be used as a potable water pipe.
- Water used for any purpose should not be returned to the potable water supply; this water should be transported to a drainage system.

FIGURE 17.12 Some hard-line facts.

Special wastes are those wastes that might have a harming effect on a plumbing system or the waste-disposal system. Possible locations for special waste piping might include photographic labs, hospitals, or buildings where chemicals or other potentially dangerous wastes are dispersed. Small, personal-type photo darkrooms do not generally fall under the scrutiny of these regulations. Buildings that are considered to have a need for special-wastes plumbing are often required to have two plumbing systems: one system for normal sanitary discharge and a separate system for the special wastes. Before many special wastes are allowed to enter a sanitary drainage system, the wastes must be neutralized, diluted, or otherwise treated.

Depending upon the nature of the special wastes, special materials might be required. When you venture into the plumbing of special wastes, it is always best to consult the local code officer before proceeding with your work.

FIGURE 17.13 Special wastes.

When you install horizontal drainage piping, you must install it so that it falls toward the waste-disposal site. A typical grade for drainage pipe is ¼ of an inch of fall per foot. This means the lower end of a 20-foot piece of pipe would be 5 inches lower than the upper end, when properly installed. While the ¼-inch-to-the-foot grade is typical, it is not the only acceptable grade for all pipes.

It you are working with pipe that has a diameter of 2½ inches, or less, the minimum grade for the pipe is ¼ of an inch to the foot. Pipes with diameters between 3 and 6 inches are allowed a minimum grade of ⅛ of an inch to the foot. Zone one requires special permission to be granted prior to installing pipe with a ⅛-of-an-inch-to-the-foot grade. In zone three, pipes with diameters of 8 inches or more, an acceptable grade is ¹⁄₁₆ of an inch to the foot.

FIGURE 17.14 Grading your pipe.

Plumbing fixtures are regulated and must have smooth surfaces. These surfaces must be impervious. All fixtures must be in good working order and may not contain any hidden surfaces that might foul or become contaminated.

FIGURE 17.15 Fixtures.

When hot-water pipe is installed, it is often expected to maintain the temperature of its hot water for a distance of up to 100 feet from the fixture it serves. If the distance between the hot-water source and the fixture being served is more than 100 feet, a recirculating system is frequently required. When a recirculating system is not appropriate, other means might be used to maintain water temperature. These means could include insulation or heating tapes. Check with your local code officer for approved alternates to a recirculating system, if necessary.

If a circulator pump is used on a recirculating line, the pump must be equipped with a cut-off switch. The switch may operate manually or automatically.

FIGURE 17.16 Hot-water installations.

Probable cause	Remedies
Excessive temperature setting	Lower setting
Defective valve	Replace

FIGURE 17.17 Troubleshooting water heaters: Relief valve drips.

Probable cause	Remedies
Rusting of inner tank walls causing pin holes	Replace heater (any repair is only temporary)

FIGURE 17.18 High operating costs with water heaters: Water tank leaks.

Probable cause	Remedies
Sediment, rust, or lime in tank	Drain and flush (replace heater if build-up is severe)
Water heater too small for job	Replace with larger heater
Wrong size piping connections	Install correct size piping
Lack of insulation on long runs of pipe	Insulate pipes

FIGURE 17.19 Troubleshooting water heaters: High operating costs.

Probable cause	Remedies
Heater installed in closed or confined area	Vent room or use louvers to allow air circulation

FIGURE 17.20   Troubleshooting water heaters: Condensation.

Probable cause	Remedies
Buildup of rust in heater	Drain and flush; if problem is severe, replace heater

FIGURE 17.21   Troubleshooting water heaters: Rusty hot water.

Probable cause	Remedies
Blown fuses or defective circuit breaker	Replace or adjust
Defective heating element	Replace
Broken thermostat	Replace
Broken time clock	Replace

FIGURE 17.22   Troubleshooting electric water heaters: No hot water.

Probable cause	Remedies
Defective thermostat	Replace

FIGURE 17.23 Troubleshooting electric water heaters: Extremely hot water comes out of faucet.

Probable cause	Remedies
Defective upper or lower heating elements	Replace defective parts
Defective thermostat	Replace
Rust buildup	Drain and clean heater

FIGURE 17.24   Troubleshooting electric water heaters: Hot water turns cold too quickly.

Probable cause	Remedies
No gas coming in	Check valves and adjust or replace
Defective thermocoupling	Replace
Dip tube installed incorrectly	Adjust—tube must be in the cold water supply

FIGURE 17.25 Troubleshooting gas-fired water heaters.

Probable cause	Remedies
Dip tube in wrong inlet or broken	Insert tube in cold water supply
Gas controller defective	Replace defective parts

FIGURE 17.26 Troubleshooting gas-fired water heaters: Water turns cold too quickly.

Probable cause	Remedies
Hole in fittings, valves, flue, or connections	Replace any worn or defective parts; tighten connections

FIGURE 17.27 Troubleshooting gas-fired water heaters: Gas smell.

Probable cause	Remedies
Defective gas control valve	Replace

FIGURE 17.28 Troubleshooting gas-fired water heaters: Extremely hot water.

	No water pressure	No water pressure at fixture	Low water pressure to fixture
Street water main	X		X
Curb stop	X		X
Water service	X		X
Branches		X	X
Valves	X	X	X
Stems, washers (hot and cold)		X	X
Aerator		X	X
Water meter	X	X	X

FIGURE 17.29 Where to look for causes of water-pressure problems.

Nominal pipe size in inches	Outside diameter in inches	Inside diameter in inches	Weight per linear foot in pounds
¼	.375	.325	.107
⅜	.500	.450	.145
½	.625	.569	1.204
⅝	.750	.690	.263
¾	.875	.811	.328
1	1.125	1.055	.465
1¼	1.375	1.291	.682
1½	1.625	1.527	.940
2	2.125	2.009	1.460
2½	2.625	2.495	2.030
3	3.125	2.981	2.680
3½	3.625	3.459	3.580
4	4.125	3.935	4.660
5	5.125	4.907	6.660
6	6.125	5.881	8.920
8	8.125	7.785	16.480
10	10.125	9.701	25.590
12	12.125	11.617	36.710

FIGURE 17.30 Facts about copper Type M tubing.

Nominal pipe size in inches	Outside diameter in inches	Inside diameter in inches	Weight per linear foot in pounds
¼	.375	.315	.126
⅜	.500	.430	.198
½	.625	.545	.285
⅝	.750	.666	.362
¾	.875	.785	.455
1	1.125	1.025	.655
1¼	1.375	1.265	.884
1½	1.625	1.505	1.111
2	2.125	1.985	1.750
2½	2.625	2.465	2.480
3	3.125	2.945	3.333
3½	3.625	3.425	4.290
4	4.125	3.905	5.382
5	5.125	4.875	7.611
6	6.125	5.845	10.201
8	8.125	7.725	19.301
10	10.125	9.625	30.060
12	12.125	11.565	40.390

FIGURE 17.31 Facts about copper Type L tubing.

Nominal pipe size in inches	Outside diameter in inches	Inside diameter in inches	Weight per linear foot in pounds
¼	.375	.305	.145
⅜	.500	.402	.269
½	.625	.527	.344
⅝	.750	.652	.418
¾	.875	.745	.641
1	1.125	.995	.839
1¼	1.375	1.245	1.040
1½	1.625	1.481	1.360
2	2.125	1.959	2.060
2½	2.625	2.435	2.932
3	3.125	2.907	4.000
3½	3.625	3.385	5.122
4	4.125	3.857	6.511
5	5.125	4.805	9.672
6	6.125	5.741	13.912
8	8.125	7.583	25.900
10	10.125	9.449	40.322
12	12.125	11.315	57.802

FIGURE 17.32 Facts about copper Type K tubing.

Length (feet)	Temperature change						
	40	50	60	70	80	90	100
20	.278	.348	.418	.487	.557	.626	.696
40	.557	.696	.835	.974	1.114	1.235	1.392
60	.835	1.044	1.253	1.462	1.670	1.879	2.088
80	1.134	1.392	1.670	1.879	2.227	2.506	2.784
100	1.192	1.740	2.088	2.436	2.784	3.132	3.480

FIGURE 17.33 Thermal expansion of PVC-DWV.

Length (feet)	Temperature change						
	40	**50**	**60**	**70**	**80**	**90**	**100**
20	.536	.670	.804	.938	1.072	1.206	1.340
40	1.070	1.340	1.610	1.880	2.050	2.420	2.690
60	1.609	2.010	2.410	2.820	3.220	3.620	4.020
80	2.143	2.680	3.220	3.760	4.290	4.830	5.360
100	2.680	3.350	4.020	4.700	5.360	6.030	6.700

FIGURE 17.34 Thermal expansion of all pipes (except PVC-DWV).

Diameter	Service weight	Extra heavy weight
2"	11	14
3"	17	26
4"	23	33

FIGURE 17.35 Weight of double-hub cast-iron pipe (30-inch length).

Diameter	Service weight	Extra heavy weight
2"	38	43
3"	56	83
4"	75	108
5"	98	133
6"	124	160
8"	185	265
10"	270	400

FIGURE 17.36 Weight of single-hub cast-iron pipe (10-foot length).

Size	Lbs. per ft.
2"	4
3"	6
4"	9
5"	12
6"	15
7"	20
8"	25

FIGURE 17.37 Weight of cast-iron soil pipe.

Diameter	Service weight	Extra heavy weight
2"	21	26
3"	31	47
4"	42	63
5"	54	78
6"	68	100
8"	105	157
10"	150	225

FIGURE 17.38 Weight of double-hub cast-iron pipe (5-foot length).

Diameter	Service weight	Extra heavy weight
2"	20	25
3"	30	45
4"	40	60
5"	52	75
6"	65	95
8"	100	150
10"	145	215

FIGURE 17.39 Weight of single-hub cast-iron pipe (5-foot length).

Diameter	Weight
1½"	27
2"	38
3"	54
4"	74
5"	95
6"	118
8"	180

FIGURE 17.40 Weight of no-hub cast-iron pipe (10-foot length).

Diameter	Weight
2"	5
3"	9
4"	12
5"	15
6"	19
8"	30
10"	43
12"	54
15"	75

FIGURE 17.41 Weight of extra-heavy cast-iron soil pipe.

Temperature (degrees F.)	Wrought iron
0	0
20	.15
40	.30
60	.45
80	.60
100	.80
120	.95
140	1.15
160	1.35
180	1.50
200	1.65
220	1.85
240	2.05
260	2.20
280	2.40
300	2.60
320	2.80
340	3.05
360	3.25
380	3.45
400	3.65
420	3.90
440	4.20
460	4.45
480	4.70
500	4.90

FIGURE 17.42 Wrought-iron steam pipe expansion (inches increase per 100 feet).

Temperature (degrees F.)	Steel
0	0
20	.15
40	.30
60	.45
80	.60
100	.75
120	.90
140	1.10
160	1.25
180	1.45
200	1.60
220	1.80
240	2.00
260	2.15
280	2.35
300	2.50
320	2.70
340	2.90
360	3.05
380	3.25
400	3.45
420	3.70
440	3.95
460	4.20
480	4.45
500	4.70

FIGURE 17.43 Steel steam pipe expansion (inches increase per 100 feet).

Temperature (degrees F.)	Cast iron
0	0
20	.10
40	.25
60	.40
80	.55
100	.70
120	.85
140	1.00
160	1.15
180	1.30
200	1.50
220	1.65
240	1.80
260	1.95
280	2.15
300	2.35
320	2.50
340	2.70
360	2.90
380	3.10
400	3.30
420	3.50
440	3.75
460	4.00
480	4.25
500	4.45

FIGURE 17.44 Cast-iron steam pipe expansion (inches increase per 100 feet).

Term	Definition
Abstract number	A number that does not refer to any particular object.
Acute triangle	A triangle in which each of the three angles is less than 90 degrees.
Altitude of a triangle	A line drawn perpendicular to the base from the angle opposite.
Angle	The difference in direction of two lines proceeding from the same point called the vertex.
Area	The surface included within the lines that bound a figure.
Arithmetic	The science of numbers and the art of computation.
Base of a triangle	The side on which a triangle is supposed to stand.
Board measure	A unit for measuring lumber, the volume of a board 12 inches wide, 1 foot long, and 1 inch thick.
Circle	A plane figure bounded by a curved line, called the circumference, every point of which is equally distant from a point within, called the center.
Complex fraction	A fraction whose numerator or denominator is itself a fraction.
Cone	A body having a circular base and a convex surface that tapers uniformly to the vertex.
Cubic measure	A measure of volume involving three dimensions: length, width, and thickness (depth).
Cylinder	A body bounded by a uniformly curved surface, its ends being equal and parallel circles.
Decimal scale	A scale in which the order of progression is uniformly 10.
Diameter of a circle	A line that passes through the center of a circle and is terminated at both ends by the circumference.
Diameter of a sphere	A straight line that passes through the center of the sphere and is terminated at both ends by its surface.
Equilateral triangle	A triangle that has all its sides equal.
Even number	A number that can be exactly divided by two.
Exact divisor of a number	A whole number that will divide a number without leaving a remainder.
Factors	Two or more quantities that, when multiplied, produce a given quantity.

FIGURE 17.45 Plumbing definitions.

FIGURE 17.45 Plumbing definitions *continued.*

Term	Definition
Factors of a number	Numbers that, when multiplied, make that specific number.
Fraction	A number that expresses part of a whole thing or quantity.
Geometry	The branch of mathematics that treats space and its relations.
Greatest common divisor	The greatest number that will exactly divide two or more numbers.
Hypotenuse of a right triangle	The side opposite the right angle.
Improper fraction	A fraction in which the numerator equals or exceeds the denominator.
Isosceles triangle	A triangle that has two of its sides equal.
Least common multiple	The lowest number that is exactly divisible by two or more numbers.
Measure	The extent, quantity, capacity, volume, or dimensions ascertained by some fixed standard.
Mensuration	The process of measuring.
Number	A unit or collection of units.
Odd number	A number that cannot be exactly divided by two.
Parallelogram	A quadrilateral that has opposite sides that are parallel.
Percentage	The rate per hundred.
Perimeter	The distance around a figure; the sum of the sides of a figure.
Perpendicular of a right triangle	The side that forms a right angle to the base.
Proper fraction	A fraction in which the numerator is less than the denomiantor.
Pyramid	A body having for its base a polygon and for its other sides of facets—three or more triangles that terminate in a common point called the vertex.
Quantity	An aspect that can be increased, diminished, or measured.
Radius of a circle	A line extending from the center of a circle to any point on the surface.
Rectangle	A parallelogram in which all angles are right angles.
Right triangle	A triangle that has a right angle (90 degree).

Abbreviation	Meaning
A or a	Area
A.W.G.	American wire gauge
bbl.	Barrels
B or b	Breadth
bhp	Brake horse power
B.M.	Board measure
Btu	British thermal units
B.W.G.	Birmingham wire gauge
B & S	Brown and Sharpe wire gauge (American wire gauge)
C of g	Center gravity
cond.	Condensing
cu.	Cubic
cyl.	Cylinder
D or d	Depth or diameter
evap.	Evaporation
F	Coefficient of friction; Fahrenheit
F or f	Force or factor of safety
ft. lbs.	Foot pounds
gals.	Gallons
H or h	Height or head of water
HP	Horsepower
IHP	Indiated horsepower
L or l	Length
lbs.	Pounds
lbs. per sq. in.	Pounds per square inch
o.d.	Outside diameter (pipes)
oz.	Ounces
pt.	Pint
P or p	Pressure or load
psi	Pounds per square inch
R or r	Radius
rpm	Revolutions per minute
sq. ft.	Square foot
sq. in.	Square inch

FIGURE 17.46 Abbreviations.

FIGURE 17.46 Abbreviations *continued.*

Abbreviation	Meaning
sq. yd.	Square yard
T or t	Thickness or temperature
temp.	Temperature
V or v	Velocity
vol.	Volume
W or w	Weight
W. I.	Wrought iron

Abbreviation	Meaning
ABS	Acrylonitrile-butadiene-styrene
AGA	American Gas Association
AWWA	American Water Works Assocation
BOCA	Building Officials Conference of America
B&S	Bell and spigot (cast-iron pipe)
BT	Bathtub
C-to-C	Center-to-center
CI	Cast iron
CISP	Cast-iron soil pipe
CISPI	Cast Iron Soil Pipe Institute
CO	Clean out
CPVC	Chlorinated polyvinyl chloride
CW	Cold water
DF	Drinking fountain
DWG	Drawing
DWV	Drainage, waste, and vent system
EWC	Electric water cooler
FG	Finish grade
FPT	Female pipe thread
FS	Federal specifications
FTG	Fitting
FU	Fixture unit
GALV	Galvanized
GPD	Gallons per day

FIGURE 17.47 Commonly used abbreviations.

FIGURE 17.47 Commonly used abbreviations *continued.*

Abbreviation	Meaning
GPM	Gallons per minute
HWH	Hot water heater
ID	Inside diameter
IPS	Iron pipe size
KS	Kitchen sink
LAV	Lavatory
LT	Laundry tray
MAX	Maximum
MCA	Mechanical Contractors Association
MGD	Million gallons per day
MI	Malleable iron
MIN (min.)	Minute or minimum
MPT	Male pipe thread
MS	Mild steel
M TYPE	Lightest type of rigid copper pipe
NAPHCC	National Association of Plumbing, Heating, and Cooling Contractors
NBFU	National Board of Fire Underwriters
NBS	National Bureau of Standards
NPS	Nominal pipe size
NFPA	National Fire Protection Association
OC	On center
OD	Outside diameter
SAN	Sanitary
SHWR	Shower
SV	Service
S & W	Soil and waste
SS	Service sink
STD. (std.)	Standard
VAN	Vanity
VTR	Vent through roof
W	Waste
WC	Water closet
WH	Wall hydrant
WM	Washing machine
XH	Extra heavy

Type of establishment	Gallons
Schools (toilet and lavatories only)	15 per day per person
Schools (with above plus cafeteria)	25 per day per person
Schools (with above plus cafeteria and showers)	35 per day per person
Day workers at schools and offices	15 per day per person
Day camps	25 per day per person
Trailer parks or tourist camps (with built-in bath)	50 per day per person
Trailer parks or tourist camps (with central bathhouse)	35 per day per person
Work or construction camps	50 per day per person
Public picnic parks (toilet wastes only)	5 per day per person
Public picnic parks (bathhouse, showers, and flush toilets)	10 per day per person
Swimming pools and beaches	10 per day per person
Country clubs	25 per locker
Luxury residences and estates	150 per day per person
Rooming houses	40 per day per person
Boarding houses	50 per day per person
Hotels (with connecting baths)	50 per day per person
Hotels (with private baths, 2 persons per room)	100 per day per person
Factories (gallons per person per shift—exclusive of industrial wastes)	25 per day per person
Nursing homes	75 per day per person
General hospitals	150 per day per person
Public institutions (other than hospitals)	100 per day per person
Restaurants (toilet and kitchen wastes per unit of serving capacity)	25 per day per person
Kitchen wastes from hotels, camps, boarding houses, etc. serving 3 meals per day	10 per day per person
Motels	50 per bed space
Motels with bath, toilet and kitchen wastes	60 per bed space
Drive in theaters	5 per car space
Stores	400 per toilet room
Service stations	10 per vehicle served
Airports	3 to 5 per passenger
Assembly halls	2 per seat

FIGURE 17.48 Potential sewage flows according to type of establishment.

FIGURE 17.48 Potential sewage flows according to type of establishment *continued.*

Type of establishment	Gallons
Bowling alleys	75 per lane
Churches (small)	3 to 5 per sanctuary seat
Churches (large with kitchens)	5 to 7 per sanctuary seat
Dance halls	2 per day per person
Laundries (coin-operated)	400 per machine
Service stations	1000 (first bay) 500 (each add. bay)
Subdivision or individual homes	75 per day per person
Marinas:	
Flush toilets	36 per fixture per hr
Urinals	10 per fixture per hr
Wash basins	15 per fixture per hr
Showers	150 per fixture per hr

CHAPTER 18
ELECTRICAL SYSTEMS

Electrical work is, in many ways, very similar to plumbing. It's not that the work itself is the same, but the risks that each phase presents a remodeler or general contractor with are closely related. Many remodeling jobs involve plumbing, but almost every remodeling job involves some form of electrical work. This is a phase of work that is hard to avoid. Builders of new construction are constantly faced with electrical installations.

Electricians are required to be licensed in order to practice their trade. This licensing requirement prohibits average people, like most contractors, from doing their own electrical wiring. This puts pressure on you to find good subcontractors in the electrical field.

How can you tell if your potential electrician is any good? References from other jobs help, and the fact that the trade is licensed doesn't hurt. However, the best way to tell is to know something about the trade. I feel strongly that it pays to put yourself in the shoes of each trade you will be dealing with. If you don't, telling the difference between a good subcontractor and a bad one is much more difficult.

There are very few remodeling or building jobs that you can take on that don't involve some electrical work. The work might be as simple as adding a ground-fault outlet or as complicated as wiring a new addition. In either case, you are going to have to rely on a licensed electrician. It is also likely that homeowners will ask you some pointed questions on the subject of electricity before you complete your estimate inspection. What kinds of questions are you likely to face? Well, let's see.

ELECTRICAL CONCERNS

Electrical concerns are likely to come up in just about any discussion of major building or remodeling. Customers will want to know what their electrical needs will be. Some of the most important questions might never be asked. Why is this? Simply because consumers are not tuned into electrical codes, so they don't know what questions to ask. They might assume many things that are not correct. It is your job, and that of your electrician, to educate these people. Let's start this discussion with some of the questions I've been asked most frequently by customers.

How Hard Is It to Change?

How hard is it to change my electrical service from fuses to circuit breakers (Figure 18.1)? This is a common question when working with old houses. While updating an electrical service should take less than a day, the job is expensive. It is also usually very worthwhile, and in some cases, it is mandatory for the remodeling scheduled to be done. A good electrician can swap out a service without much disruption to the homeowner. The power will be off for several hours, but the job will not take more than a day to complete.

Can I Use Electric Heat in My New Space?

Customers who are cost-conscious on their initial cost of remodeling frequently explore using electric heat, because it is the cheapest heating option normally available. Electric heat requires a lot of electricity. Houses that have old 60-amp fuse boxes cannot have electric heat installed unless the electrical service is upgraded. Homes with 100-amp services and circuit breakers are limited in what amount of electric heat, if any, they can support.

Standard procedure calls for a 200-amp electrical service and circuit breakers to be used in conjunction with electric baseboard heat. Many new houses have 200-amp services, but there are an awful lot of houses being lived in that don't have adequate electrical services to work with electric heat.

What Are These Ground-fault Things?

People have heard about ground-fault interceptors (GFI), but most of the folks don't know what these devices are, when they are needed, or how they work. I assume that you are aware of what GFI devices are, but let me expand on this subject.

- 10 feet above ground
- 10 feet above walkways
- 15 feet above residential driveways
- 18 feet above public streets

FIGURE 18.1 Minimum clearances for service drops.

GFI devices can be in the form of circuit breakers or individual outlets. The outlets are much cheaper. These devices are installed in "wet" areas. The purpose of a GFI device is to protect people from hazardous electrical shocks. Because water and electricity don't make a good match in terms of personal safety, GFI breakers or outlets are required in certain locations. These locations typically included the following:

- Bathrooms
- Kitchens
- Outside outlet locations
- Darkrooms
- Laundry Rooms (where a sink is installed)
- Garages

Basically, a GFI protection device should be installed in any location where a person might be in contact with water and electricity at the same time. For example, a whirlpool bathtub should be wired into a GFI-protected circuit. This would be done with a GFI circuit breaker. Breakers can be used for all GFI circuit locations, but GFI outlets are a less expensive option. By having the first outlet in the circuit be a GFI outlet, all remaining outlets on the circuit are GFI protected. The electrical code requires GFI protection in locations like those just mentioned, so GFI protection can't be considered an option; it is a necessity.

How Will You Get the Wires up There?

When people are considering attic conversions, they often ask how you will get the wires up there. It might be possible to fish the wires through existing walls, but this cannot be counted upon. Because most attic conversions require heat and plumbing, it is sensible to assume that some chase ways will have to be made or that some existing walls will have to be opened up. This provides a path for electrical wires, as well.

THE UNASKED QUESTIONS

The unasked questions pertaining to electrical work are often more important than the ones that are presented. Most homeowners don't know enough about electrical systems to understand what questions to pose. This is where your knowledge and expertise shines. If you can point out key issues to potential customers before they are even aware that such issues exist, you have a much better chance of making a sale. With this in mind, let's peruse some of the unasked questions that you can capitalize on.

Is Your Electrical Service Adequate for All the Changes?

This is a question you should ask whenever a customer is requesting the addition of new circuits. A garbage disposer or dishwasher might seem innocent enough; however, if there is no room left in the panel box for a new circuit, an easy job can turn into an expensive mistake. You should always check the electrical panel before you make any firm commitments to providing additional circuits.

Are You Aware of the Code Requirements?

This question should always be asked. Honest homeowners will almost always say that they are not aware of code requirements. This is your chance to show off a little and to impress the homeowner.

For example, you can quote the need for GFI protection. This can be followed with code requirements on outlet spacing. Essentially, wall outlets cannot be more than 12 feet apart. The code requires a light or appliance with a 6-foot electrical cord to be placed in any location along a wall without having to use an extension cord to reach an outlet. Along kitchen counters, the spacing is reduced to 4 feet. These types of statistics impress people, and if you can impress them, you can sell them. However, don't take the code regulations that I'm giving you here for gospel, check your local code for current requirements.

Have You Thought about Your Switch Locations?

Have you thought about your switch locations and which switches will control which lights? This question can be very important during the planning stage of a job. By showing people how they can specify the location of switches, within reason, and how they can suggest what lights the switches operate, you can give the customer a better job. People pick up on these types of questions, and they identify them with caring and concern on your part. This is a big step towards closing a sale.

Have You Set a Budget for Your Light Fixtures?

Most electricians bid their jobs without including the price of light fixtures. They will typically detail a fixture allowance in their bid, but their price might not include any allowance for fixtures. If you forget to mention this fact to your customers, someone might have to come up with hundreds of dollars that they hadn't planned on. One fixture, such as a dining room chandelier, can cost several hundred dollars. It is also possible to buy fixtures for less than $10. Before you can bid a job successfully, you must establish a lighting allowance with your customer.

This list of questions could continue for several more pages, but you should be getting the idea of what I'm telling you. Don't wait for your potential customers to ask all of the questions. It is not fair for you to expect them to know what questions to ask. You should prompt them with questions that are pertinent to the job. In doing this, you head off problems before they develop, you win more jobs, and you have happier customers.

EXISTING CONDITIONS

Existing conditions with electrical remodeling do not normally create as many problems for a remodeler as plumbing can. This is not to say that existing electrical systems can't cause trouble for you. The primary concern for contractors in most remodeling jobs, pertaining to electrical work, is the electrical service. If the panel box is in compliance with code requirements, has adequate room for any additional circuits being created, and is in satisfactory condition, the remainder of existing wiring does not necessarily affect you. Unlike plumbing, where old pipes are being tapped into, old wiring is left alone during remodeling. By this, I mean that new circuits are not tied into existing circuits. Naturally, your electrician might have to reroute existing wires. This can get you involved in existing conditions. However, if you are doing an attic conversion, a basement conversion, or building an addition, you should not have any reason to work with old wiring. All of your work will be focused on running new wires to the existing panel box.

Because the panel box is so critical, this is one of the first things that you should look at during your estimate inspection. If the service is a fuse box, you can expect some trouble. When the box is a 100-amp circuit-breaker box, you should still expect some trouble. With today's houses, a 100-amp electrical service is considered small. If you are doing any significant additional wiring, a 100-amp box might not be sufficient.

A 200-amp service will generally be large enough to avoid major electrical upgrades in the service panel, but you can't just assume this. If a house is large or has a number of separate circuits, even a 200-amp panel can become full quickly. If you look in a panel box and see that it is full, or nearly full, you should be aware that costly modification to the electrical service for the home might be needed before your work can be completed (Figure 18.2). This is a good time to have your electrician look over the job to give you firm price quotes.

I don't want to give you the impression that the electrical service is the only place where existing electrical conditions can affect you. It is not. For an example, you might be replacing an existing water heater as a part of your remodeling job. This is certainly not unusual. Neither is it unusual for older electric water heaters to be connected to wires that are, by present code requirements, too small. It was common for years to run 12-gauge wire to a water heater. Under today's code requirements, the wire must be no less than 10-gauge. The current code also requires an independent disconnect box to be located near the water heater. This has not always been the case. So, if you stumble onto a water heater that is wired with 12-gauge wire and no disconnect box, your cost to replace the heater will have to include upgrading the

Device	Wattage rating
Room air conditioner (7000 BTU)	800
Clothes dryer	5800
Clothers washer	600
Furnace (blower)	1000
Furnace (oil burner)	300
Humidifier	80
Sewing machine	90
Television	120
Vacuum cleaner	650
Water heater	4500
Shallow-well pump	660
Deep-well pump	1320

FIGURE 18.2 Common wattage ratings for general household devices.

wire and installing the disconnect. A job like this can get expensive, and it can eat away at your anticipated profit.

Kitchens

Are you aware that major kitchen appliances are required to have their own electrical circuit? Well, they are, and many older homes were not wired in this manner. If you are doing extensive kitchen remodeling, there are several code upgrades that might be needed (Figure 18.3). Your electrician might have to run a whole new circuit for kitchen outlets. Kitchen outlets are required to be installed on two separate circuits. GFI protection will be required in the kitchen, and appliance wiring might have to be reworked to provide individual circuits. All these new circuits can fill a small panel box quickly, so you have to be able to evaluate these existing conditions during your estimate phase.

Bathrooms

Let's talk about bathrooms for a moment. If you are gutting and redoing a bathroom, there are a couple of electrical issues to be aware of (Figure 18.4). A bathroom is required to have GFI protection. Many old bathrooms don't have this, so you will have to plan on the expense to provide it. If the bathroom does not have an operable window, a ventilation fan will be required.

This will cause you to run new wiring, buy and install the fixture, and extend a vent hose for the fan. Finding a way to vent a fan that is located in a bathroom can be tricky. If the bathroom doesn't have attic space above it or isn't built on an outside wall, getting the vent to open air space can require extensive work. If you fail to pick up on this during your estimate, the cost for doing the job might come out of your pocket.

When you remodel a bathroom, your work might very well include the installation of a whirlpool tub. If it does, you will need to wire the motor for this fixture with a GRI circuit. This isn't particularly difficult, but a GFI breaker costs about 10 times what a regular breaker does, and then there is the cost for more wire than you might have thought would be needed. On top of this, there is the labor for a licensed electrician to run the wire and install the breaker. These are both little jobs that you might not have figured on. All in all, not knowing that a whirlpool needs to be GFI protected can cost you a few hundred dollars.

Device	Wattage rating
Blender	375
Coffee maker	1000
Dishwasher	1000
Freezer	500
Frying pan	1200
Microwave oven	1200 to 1800
Range	12,000
Refrigerator	350
Toaster oven	1200

FIGURE 18.3 Common wattage ratings for electrical devices in kitchens.

Device	Wattage rating
Bath fan	100
Hair curler	1200
Hair dryer	1200
Heater	1500
Heat lamp	250

FIGURE 18.4 Common wattage ratings for electrical devices in bathrooms.

NEW INSTALLATIONS

When you become involved with new installations of electrical wiring and fixtures, the work will be subject to local code requirements (Figures 18.5 through 18.9). This should not cause you any specific trouble. A qualified electrician should bid all jobs so that the work will comply with local code requirements. There are aspects of new electrical installations that you should discuss with your customers to avoid conflict and confusion later on. Let me give you some examples.

A Ceiling Fan

A ceiling fan is a popular add-on during remodeling jobs. Ceiling fans can be purchased at very affordable prices, but the cost to install them can escalate

Code	Meaning
H	Heat-resistant
R	Rubber
T	Thermoplastic
W	Water-resistant
AC	Armored cable
C	Corrosion-resistant
F	Feeder
NM	Nonmetallic
U	Underground
SE	Service entrance cable

FIGURE 18.5   Codes that identify wiring insulation.

Wire color	Status	Connects to . . .
Black	Hot	Darkest screw on device
Red	Hot	Second hot wire in a 240-volt circuit
White	Neutral*	Silver screw on device
Green	Ground	Green screw on device

Note: White wires are sometimes used as additional "hot" wires. Most electricians color the white wire black to indicate this type of use, but don't bet your life on it. Take meter readings on all wires before touching them.

FIGURE 18.6   Color codes used in electrical installations.

Outside dimension (inches)	Wire size #6	Wire size #8	Wire size #10	Wire size #12	Wire size #14
$2 \times 3 \times 2\frac{1}{4}$		3	4	4	5
$2 \times 3 \times 2\frac{1}{2}$		4	5	5	6
$2 \times 3 \times 2\frac{3}{4}$		4	5	6	7
$2 \times 3 \times 3\frac{1}{2}$	3	6	7	8	9

FIGURE 18.7 Number of conductors allowed in rectangular boxes.

Outside dimension (inches)	Wire size #6	Wire size #8	Wire size #10	Wire size #12	Wire size #14
$4 \times 1\frac{1}{4}$		4	5	5	6
$2 \times 1\frac{1}{2}$		5	6	6	7
$2 \times 2\frac{1}{8}$	4	7	8	9	10

FIGURE 18.8 Number of conductors allowed in round and octagonal boxes.

Outside dimension (inches)	Wire size #6	Wire size #8	Wire size #10	Wire size #12	Wire size #14
$4 \times 1\frac{1}{4}$		6	7	8	9
$4 \times 1\frac{1}{2}$	4	7	8	9	10
$4 \times 2\frac{1}{8}$	6	10	12	13	15
$4\frac{11}{16} \times 1\frac{1}{4}$	5	8	10	11	12
$4\frac{11}{16} \times 1\frac{1}{2}$	5	9	11	13	14

FIGURE 18.9 Number of conductors allowed in square boxes.

quickly. Let's say, for example, that you have been hired to build a sun room. The sun room will connect to an existing living room. As a part of this improvement, the customer wants to have a ceiling fan installed in the living room. The living room has a vaulted ceiling and a ceiling fan will add to the appearance and function of the room. You are doing your estimate for the sun room when the homeowner comes out of left field with the request for a ceiling fan in the living room. You jot down the request in your notes and nod your head affirmatively. Now what happens?

You might have just put yourself into a bind. How are you going to get a support bar and box installed in the living room ceiling without damaging the ceiling? You aren't going to. The ceiling will have to be cut to allow the box and bar to be installed properly. Now, who's going to assume responsibility for patching and painting the damaged area? Is the new paint going to match the paint on the rest of the ceiling? If it doesn't, are you going to pay to have the entire ceiling painted? How are you going to wire the switch for the fan, without damaging the living room wall? Remember that you are working with a vaulted ceiling where there is no attic. This means that routing the wires will be most difficult. Can you see now how a little add-on can make your life miserable?

In the example you've just been given, you are lucky. Because the sun room is being built so that it adjoins with the living room, you have some latitude to move about that you would not have if the sun room wasn't being built. The common wall between the sun room and the living room gives you an opportunity to route your wires without destroying the living room wall. Even with the benefit of having open walls to work with in the sun room, adding this simple, inexpensive ceiling fan can wind up costing you hundreds of dollars.

Ceiling fans are not the only new fixtures that can cause trouble similar to that which we have just discussed. Fans installed with support bars are more difficult to deal with than simple light fixtures, but adding any new fixture can create difficulties for an electrician. Before you agree to do any new installations for a customer, you should check for potential problems carefully. Ideally, you should have your electrician available to make judgment calls on potential complications.

Access

Access for new wiring is not usually a big problem. Most homes have attics, basements, or crawl spaces where wires can be installed. Some of these locations are easier than others to work in. For example, a basement provides much better working conditions than a crawl space. An electrician can work faster in a basement than in a crawl space. If your electrician can work faster, the cost of labor should be less. Conversely, if you spot areas where access will be difficult, you should plan on higher labor costs. This applies not only to electrical work, but to all types of work.

People often assume that electrical wires can be pulled through existing walls without damaging the finish on the walls. This is, indeed, true in many cases, but not in all. If a wall has fire blocking installed in it, getting a wire up or down the wall without opening the wall will not be possible. If you assure a customer that your electrician can snake wires through a wall, you'd better hope that the wall doesn't contain obstructions that prohibit the pulling of wires. I feel that you should refrain from telling customers information that you cannot be sure of. If you turn out to be incorrect in your information, you will lose credibility and you might have some angry customers on your hands.

Garbage Disposers

Garbage disposers and dishwashers are common appliances added during kitchen remodeling. Wiring these appliances is not normally a big deal. An electrician can usually gain access to a panel box, insert a breaker, snake a wire under the house, and run the wire to the location of the new appliance. Even when a house is built on a slab foundation, the electrician can usually run wires through an attic to get them to the location of a new appliance.

However, this simple type of wiring is not always possible. If the kitchen has an exposed-beam ceiling with no attic and is sitting on a concrete slab, fishing a wire to a new disposer or dishwasher can be much more difficult. You have to look for these types of obstacles before you can commit to a price for a customer.

All new installations can involve access problems. In my opinion, access and existing electrical services are the two most critical aspects for a remodeler to look for when doing an estimate that involves electrical work. I've mentioned throughout this book that you should consult with experts on any issues that you are not absolutely sure of. I maintain that this advice is critical to your success as a remodeling contractor. If you are a builder of new homes, you should have an electrical diagram, found on the blueprints for the home, to go by. In any case, you must be precise in your compliance with local electrical code requirements.

CHAPTER 19
INSULATION

Working around insulation can be an aggravating experience. Many people find insulation to be very irritating to their skin. Some contractors have trouble breathing normally when working with insulation. Yet, insulation is a part of every modern construction job. This means that you will come into contact with it at some time in your work. Not all jobs put contractors face to face with insulation, but some do. Even if you aren't involved in the installation of new insulation, you are very likely to have to put up with insulation in one form or another.

Have you ever removed a ceiling and had old insulation rain down on you? If you have, you know how uncomfortable the rest of the work day can be. Getting insulation in your face, in your clothes, and all over you can make finishing out the day a real struggle. Even if you are wise enough to check above a ceiling before you open it, you might not be able to avoid old insulation. When this is the case, the best you can do is prepare to work around it.

Installing new insulation is not a very technical type of work. Almost anyone can learn quickly how to install most forms of insulation. As a contractor, you can have one of your laborers or trainees do the dirty work for you, but this might not prove to be profitable. It is one thing to be able to install insulation, it is quite another thing to be competent enough to install it profitably.

Insulation is one phase of construction and remodeling work that I've never enjoyed. Try as I have, there seems to be no way to escape working with it. I have used helpers to do installations, but I've never made much money working in this manner. My experience has proved that, for me at least, more money can be made by hiring specialized insulation companies to do the installation work for me. I don't know how the companies make money. You see, I've found that aggressive insulators will provide the labor and material to insulate a job for about what I have to pay just for the insulation. This essentially means that I am getting the installation for free.

Unless you are working on an extremely tight budget and doing most work with your own two hands, there is little justification in doing your own insulation work. A good crew from a professional insulating company can get the job done much faster than you can, and the cost will probably not be much more than what you would spend on materials. Your time can be put to better, and more profitable, use. Go out and sell another job, that's where the real money is.

TYPES OF INSULATION

There are many types of insulation available. Some types are better suited for certain types of jobs than others. The R-values of insulation vary (Figure 19.1). While most contractors have standard procedures for their insulation installations, it can pay to know what all of your options are. As with any other aspect of your business, the better informed you are, the better off you are.

Regardless of whether you install your own insulation or sub the work out, you have to know enough about the various products to satisfy your customers. You might go for months, or even years, without having a customer ask for a detailed comparison of the types of insulation available. However, if this question comes up during a meeting between a perspective customer and yourself, you could lose a lot of credibility if you are unprepared to answer it. Let me give you a quick example of what I mean.

Let's say that you are a remodeling contractor and that I'm your potential customer. You are in my home, pitching me on the virtues of your company. During the presentation of your proposal, I notice that the quote is ambiguous on the issue of insulation. The clause in the proposal states that the job will be insulated in compliance with local codes, but it doesn't give me the kinds of

Material	R-value per inch of insulation
Fiberglass batts	3
Fiberglass blankets	3.1
Fiberglass loose-fill	3.1 to 3.3 (when poured), 2.8 to 3.8 (when blown)
Rock-wool batts	3
Rock-wool blankets	3
Rock-wool loose-fill	3 to 3.3 (when poured), 2.8 to 3.8 (when blown)
Cellulose loose-fill	3.7 to 4 (when poured), 3.1 to 4 (when blown)
Vermiculite loose-fill	2 to 2.6
Perlite loose-fill	2 to 2.7
Polystyrene rigid	4 to 5.4
Polyurethane rigid	6.7 to 8
Polyisocyanurate rigid	8

FIGURE 19.1 R-values for insulation.

details I want. After reading your entire proposal, I sit back and start asking for clarification on certain areas of the agreement, one of which has to do with insulation.

My first question pertains to the attic insulation you will be installing. Will it be in batts, or will it be loose-fill insulation. You tell me that it will be loose-fill. Then I ask if the insulation will be blown into place or installed by hand. You stumble over words, trying to think while you talk. It appears to me that you have not considered the method of installation. This concerns me. Because you don't know how the insulation will be installed, how do you know how much to charge me? Have you just pulled your prices out of thin air?

My next question requires you to describe the type of loose-fill material that will be used. Will it be a glass-fiber product, mineral wool, or cellulose. You are again at a loss for words, and this makes my impression of your professionalism dwindle. It is becoming obvious to me that you don't know much about the insulating work that will be done for me.

Even though I'm guessing that you are not up to speed on insulation, I ask you to explain the pros and cons of mineral wool insulation. I go on to ask for a detailed evaluation of how glass-fiber insulation stacks up against cellulose. Your lack of skilled responses is really starting to worry me. Should I eliminate you from consideration simply because you are not fluent in insulation details? I probably shouldn't, but I might. You could be the best remodeling contractor in the area; however, if you impress me as someone who doesn't take technical issues seriously, I might not trust you to see my job through to a successful completion.

By not knowing the answers to my questions pertaining to insulation, you could lose the entire job. The fact that you rely on professional insulation contractors to advise you on what types of insulation to use could be lost on me. If these professionals were present to answer my questions, there wouldn't be any problem. However, because you are alone and unable to give solid responses to my questions, I'm definitely going to lose some confidence in you. This is something no contractor can afford to have happen. To avoid getting boxed into a corner, you need enough knowledge of insulation to carry on a competent conversation. Now that we have established a need for you to know more about insulation, let's examine the various types available and what their prime uses are.

Glass-fiber Insulation

Glass-fiber insulation is, by far, the most widely used type of residential insulation. It is installed in crawl spaces, exterior walls, and attics. The insulation is available in various forms. You might buy loose-fill material, faced batts, unfaced batts, and so forth. R-values vary, based on the thickness of the insulation.

As well known and popular as glass-fiber insulation is, it is also one of the more difficult types to work with if you have sensitive skin. Glass-fiber insulation makes a lot of people itch. This condition is usually worst in hot weather, but it can occur at any time of the year. Aside from the irritating nature of glass-fiber insulation, the remainder of its features and benefits are basically good.

Batts. Batts and blankets of glass-fiber insulation (Figure 19.2) are used in attics and exterior wall cavities. Gaining access to an attic to install batts and blankets is usually easy; however, to insulate a wall with this type of material, the entire wall with have to be opened up. In some remodeling jobs this isn't a problem, because the structure might be having all of its interior wall-coverings removed from exterior walls. If destroying wallcoverings will be a problem, you can still use glass-fiber insulation.

Loose-fill. Glass-fiber insulation is available in a loose-fill form. This type of insulation can be spread around an attic by hand, or it can be blown into attics and exterior walls. Unlike batts, where entire walls have to be opened for installation, loose-fill material can be blown into wall cavities through small holes. The holes are much easier and less expensive to repair than a complete rip-out of the wallcoverings.

Rigid Boards. You might not expect to find glass-fiber insulation in the form of rigid boards (Figure 19.3), but you can. These boards are used to add insulating value to exterior walls that are being constructed. The rigid boards

Thickness (inches)	R-value
3½	11
3⅝	13
6½	19
7	22
9	30
13	38

FIGURE 19.2   Ratings for fiberglass batt insulation.

Widths	Lengths	Thickness
4 feet	4 feet to 16 feet	½"
		⅝"
		¾"
		1"

FIGURE 19.3   Sizes of insulation boards.

Type	R-value per inch of insulation
Cellulose	2.8 to 3.7
Fiberglass	2.2 to 4.0
Perlite	2.8
Rock wool	3.1
Vermiculite	2.2

FIGURE 19.4 Ratings for loose fill insulation.

are also used to insulate basements walls, both inside and out. They can even be used to help insulate vaulted ceilings, where there is no attic. In terms of rigid insulation, glass-fiber doesn't stack up well against it competitors: polystyrene and urethane.

Mineral Wool

Mineral wool is, in many ways, similar to glass-fiber insulation. This insulation is available in batts, blankets, and loose-fill. It's R-Value, per inch of insulation, is the same as that of glass-fiber insulation. Mineral wool is, by no means, a poor insulator; however, if you don't like to itch and you want a slightly higher R-value, cellulose might be worth considering.

Cellulose

Cellulose insulation is limited in its use. Because the product is available only as a loose-fill insulator (Figure 19.4), it is not practical to install cellulose in the stud bays of new construction. If you will be blowing insulation into existing walls, cellulose is a worthy contender. Its R-Value rating is slightly better than mineral wool and glass-fiber insulation. There is both good and bad to assess with cellulose.

If cellulose insulation gets wet, it loses much of its insulating quality. Not only this, untreated cellulose is a considerable fire hazard. When you plan to install cellulose insulation, make sure that it has been treated to be fire resistant. Old paper is the prime ingredient in cellulose insulation. Knowing the properties of paper, you can imagine how cellulose insulation performs when it is subjected to extended moisture, insects, rodents, and so forth. It doesn't fare well.

The strong points to cellulose are that it is affordable and will not normally cause irritations for installers. There is also the fact that recycled paper is what cellulose is made of, so there is little to worry about in terms of odor emissions or health threats.

Polystyrene

Polystyrene is used in the construction of rigid insulation boards. The insulating quality of polystyrene is very good. The downside to this insulator is its cost and the fact that it can be flammable. The R-value for polystyrene is the same as that of glass-fiber insulation and mineral wool. All of these insulation materials share a rating of R-3.5 for every inch of insulation installed.

Urethane

Urethane is one of the most effective insulators available, but it is illegal to use in some locations. If you consider that most insulation materials have a value of around R-3.5 and urethane has a rating of R-5.5, it is easy to see why urethane is known as the leader of R-values. Unfortunately, urethane is also known to produce cyanide gas if it burns. This, of course, is a deadly gas. Due to the potential risk of creating a poisonous gas, urethane insulation has been banned in a number of locations.

Urethane is available in the form of rigid boards and as a foam. The foam version was extremely popular for old homes made of brick and block. If allowed by local codes, urethane is far and away the most effective insulation material you can use, in terms of R-values. However, you aren't likely to fill the stud bays of a new addition with foam. You have to plot your work in accordance with your personal circumstances.

R-VALUES

R-values are a unit of measurement intended to establish the resistance of a certain material (Figure 19.5). The higher the resistance level is, the better insulating quality the material has. For example, insulation with a rating of R-19 is not as good as an R-30 insulation. Most homeowners are familiar with R-values, so you shouldn't be forced to educate many people on what R-values are or how they work. However, you might get some questions pertaining to existing building materials and their R-values.

Thickness (inches)	R-value
3½	11
3⅝	13
6½	19
7	22

FIGURE 19.5 Ratings for rock wool batt insulation.

Material	R-value per inch or as specified
Concrete	0.11
Mortar	0.20
Brick	0.20
Concrete block	1.11 for 8"
Softwood	1.25
Hardwood	0.090
Plywood	1.25
Hardboard	0.75
Glass	0.88 for single thickness
Double-pane insulated glass	1.72–⅝ w¼" air space
Air space (vertical)	1.35–¾"
Gypsum lath and plaster	0.40–⅞"
Dry wall	0.35–½"
Asphalt shingles	0.45
Wood shingles	0.95
Slate	0.05
Carpet	2.08
Vinyl flooring	0.05

FIGURE 19.6 R-values for common building materials.

Do you know what the average R-value of a single-glazed window is? I know that windows are normally rated with U-values, where the lower the rating the better the window. The R-value of an uninsulated window is about R-1. A double-glazed window should have a rating in the neighborhood of R-2. What would you guess the R-value of an average older door to be (Figure 19.6). If you guessed R-1, you should be in the ballpark. Storm doors can raise the rating to R-2.

When customers are talking with you about adding new insulation, they might want an idea of what their existing building components are doing for them in terms of R-value. A typical exterior wall, in a wood frame house, that is covered with wood siding will carry a rating for around R-5. If an insulating sheathing as been installed beneath the siding, this rating could go up to R-7. If a home has an 8-inch brick wall, the R-value will probably be around R-4.

Ceilings that are made of drywall normal carry an R-value of 4. This rating is subject to the attic conditions, if there is an attic, over the drywall ceiling. It is not unreasonable to assume an R-value of 8 in some circumstances. You can't arrive at an accurate R-value unless you know what all of the existing materials are and the circumstances surrounding them.

Type of insulation	R-value
Glass-fiber	R-3.5
Mineral wool	R-3.5
Cellulose	R-3.6
Vemiculite	R-2.2
Perlite	R-2.4
Polystyrene	R-3.5
Urethane foam	R-5.5

FIGURE 19.7 R-values for different types of insulation.

Floors made of wood might carry a rating of R-4. If carpeting is installed over the floor, this rating might hit R-6. Most houses, even old ones, will have some insulation in them. Attics and crawl spaces are the most likely areas to find this insulation in. Because access is better for an attic or floor than it is for an exterior wall, these locations usually get top billing when it comes to doing an energy upgrade. To evaluate the R-value of insulation, you must have some means of measurement. This is typically done by measuring the depth of the insulation and converting the depth to an R-value. To do this conversion, you need some numbers to plug in. They are listed in Figure 19.7 (all R-values are based on one inch of insulation).

VAPOR BARRIERS

Vapor barriers are needed when installing insulation. Without them, moisture can build up in wall cavities and cause wood products to rot. Mildew is another potential side effect. There are several ways to create a vapor barrier. Many manufacturers offer both faced and unfaced batts of insulation. The facing on a batt of insulation serves as a vapor barrier. Insulation contractors frequently install unfaced insulation and then cover it with plastic. This also creates a vapor barrier.

Condensation can be a big problem in some houses. When condensation forms, moisture is present. This moisture can cause a house to deteriorate before its time. If a proper vapor barrier is not installed, condensation can rot wood structural members and reduce the efficiency of insulation to half of its normal R-value rating. This damage generally occurs over a number of years and is not normally found until significant structural damage has been done. As a remodeler, you are in a prime position to discover this type of problem. If you are called in to install a new window or to replace some drywall, you might find that condensation has wreaked a wall. Be on the lookout for this.

Vapor barriers should be installed towards the heated space of a home. If you are using faced insulation, the facing should be visible from the living space of the house, prior to being covered with drywall. Installing insulation backwards, and I've seen jobs where this has been done, will result in some serious moisture problems.

I remember an article in a newspaper showing how an nearly new house had rotted because the vapor barriers on the outside walls had been installed backwards. Instead of repelling moisture back into the house, these backwards barriers trapped water and caused it to saturate the insulation. The result was ruined insulation, rotted wood, and a very, very unhappy customer.

From personal experience, I've found most mistakes with vapor barriers to be made in crawl spaces. This is the one location where I have personally encountered insulation installed with the vapor barrier upside down. Whether these installations were done by professional contractors or homeowners I don't know, but I do know that the jobs had been done incorrectly.

VENTILATION

Ventilation is needed in a home, and installing insulation and vapor barriers can reduce ventilation to a point where air inside the home might not be healthy. The current construction field has undergone numerous changes over the years. Some have proved to be good, and others have not enjoyed such enviable track records. One mistake learned during this time is that it is possible to make a house too tight. If air is not allowed to come and go in a house, big problems can crop up. As a contractor, you should be aware of these potential problems.

People want to conserve energy and money, so they hire people like you to tighten up their homes. Weatherstripping is added, replacement windows are installed, caulking is done, and insulation and vapor barriers are installed. If you carry this work to extremes, you might be creating a very dangerous situation. You might think you are giving your customer a perfect home, when in reality, you might be creating a nightmare of physical complications for the residents.

Are you aware that carpeting and furniture can emit dangerous substances? Do you know what Radon is and how it affects people? How much do you know about the vapors and fumes that might accumulate during an average day's cooking in a kitchen? Going to an extreme, how long can a person breathe stale air before the oxygen levels are depleted? All of these questions pertain to what we're talking about. If a house is sealed up too tightly, any of the issues we have just touched on can grow in magnitude.

How many houses get wrapped before they get sided? A lot. Would you assume that most newer houses are filled with insulation and secured with plastic vapor barriers? I would. Are today's windows and doors tighter than the ones that where in your grandparent's home when you were a toddler? They certainly are. With all the fuss to create a more efficient home, some builders have created monsters. The houses they built are too tight. Don't allow yourself to fall into this same trap. Give your customers a good job, but don't seal them in so tightly that they will suffer from the potential consequences.

CHAPTER 20
INTERIOR TRIM

Interior trim is a big part of a finished job. If the trim doesn't look good, the job doesn't look good. The most difficult part of interior trim when installing it in a remodeling job is matching it to existing trim. This can be a real problem, especially in older homes. Builders of new homes are not faced with this problem. Most homeowners recognize the fact that making exact matches between new trim and old trim is not an easy or inexpensive task. This doesn't mean that they don't want the merger to be as attractive as possible.

What type of trim do you install most often? I would guess it is colonial trim. This is, by far, the most popular type of trim in modern construction. Some jobs are done with clamshell trim, and others are done with rustic boards. The choice of which trim to use is strictly cosmetic. There is no structural value to the material, and one type of trim will hide a crack or seam as well as any other type will (Figure 20.1). So, what's the big deal with trim? Appearance. People like to live in homes that look good, and trim provides much in terms of physical appearance.

GO WITH THE FLOW

When it comes to trim, it usually pays to go with the flow. If a majority of homes in an area have colonial trim in them, it would not be wise to install clamshell trim in a house. The appraised value of the home would likely suffer. Colonial trim is so well accepted that it is hard to go wrong with it. One exception would be homes with a rustic flair. When a house is capturing a pioneer spirit, plain dimensional lumber works well as a trim. Clamshell trim and vinyl baseboard don't have much place in a competitive market. These

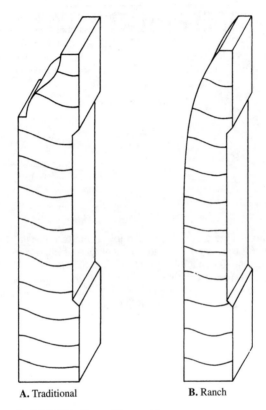

A. Traditional   **B.** Ranch

FIGURE 20.1 Types of base moldings.

materials are typically used in projects where money is the measuring stick used to make decisions. However, saving money on trim can cost money later on, such as on an appraisal.

Almost all the houses I build and remodel are fitted with colonial trim. This is not just my personal preference, it is a decision I've made based on years of experience. I've seen how people respond to various types of trim (Figures 20.2 and 20.3), and I've never had anyone complain about colonial trim. During my time in this business, I've met with a lot of real estate appraisers. These meetings and consultations have proved colonial trim to be a safe bet. For these reasons, I stick with it.

I know a contractor who swears by your basic one-by material as a finish trim. He uses it everywhere he can. I think this type of trim is fine in a rustic home, but I can't see it in a formal setting.

You must have your own preference in terms of trim, and this might drive you to lead customers to your favorite. As professionals, we have to remember that we work for our customers. It is our obligation to make them aware of their many choices, but we should not insist that they take all of our advice. You should lay out the options for your customers and allow them to choose for themselves.

FINGERJOINT TRIM

Do you use much fingerjoint trim? Some contractors swear by it, and others swear at it. I don't see anything wrong with using fingerjoint trim when it is going to be painted. If the trim is going to be stained, fingerjoint is not a good choice. However, fingerjoint is less expensive than stain-grade trim. Also, when painted, fingerjoint trim looks the same as clear trim.

- Crown
- Rabbeted half round
- Half round
- Corner bead
- Sliding door
- Handrail
- Cove
- Quarter round
- Dowel
- Picture rail
- Scoop
- Edge

FIGURE 20.2 Types of trim molding.

- Crown: Trims corner between wall and ceiling and can act as a mantle trim.
- Cove: Trims area between walls and ceilings and is used for inside corner trim.
- Quarter round: Used to conceal gaps between flooring and walls or baseboards.
- Corner bead: Trims outside corners.
- Half round: Used to hide joints between two butting pieces of paneling.

FIGURE 20.3 Uses of trim molding.

SOLID TRIM

Solid trim costs more than fingerjoint trim, but it is the best type of trim to use when stain will be applied to the wood. Some people call solid trim "clear trim," and others call it "stain-grade trim." I normally refer to it as stain-grade trim. Regardless of what you call it, solid trim doesn't have all the visible joints that exist in fingerjoint trim.

I've seen people use fingerjoint trim and stain it. I can attest to the fact that it doesn't look good. On one occasion, fingerjoint trim was used around interior doors while solid trim was used elsewhere. All of the trim was stained. The result was less than desirable. When fingerjoint trim is stained, it is not a pretty picture.

MATCHING OLD TRIM

Matching old trim with new trim can be quite a challenge. There are many times when this just isn't possible with standard, stock trim. However, if your customer is willing to pay, you can have custom trim made. It is also possible to buy replica trim from specialty suppliers. Don't expect the price to be meager. Your customers will have to be committed to making an investment in continuity. If not, you will just have to mix and match the trim.

If you happen to be a dedicated woodworker, you might make your own replica trim. If you do this, you can expect to net a tidy sum for your trouble. The money to be made in matching old trim can be staggering. However, you've got to have customers who are willing to pay dearly for their desires. This is not always the case. A lot of people don't object to mixing modern trim with authentic trim. This is a subject that you will have to work out with your customers.

METICULOUS WORK

Interior trim is meticulous work when it is installed properly (Figures 20.4 and 20.5). A sharp saw and a sharp eye are key elements to fine trim work. Cutting angles that fit together tightly is essential to a professional trim job. If you can make good cuts, you might say that you are cut out to be a trim carpenter.

Do you use air equipment? Many trim carpenters are now using pneumatic nailers. If you are in this group, you are probably familiar with your tools. However, some trim carpenters don't set their air equipment up right. They don't shoot the nails deep enough. This causes problems for the painters

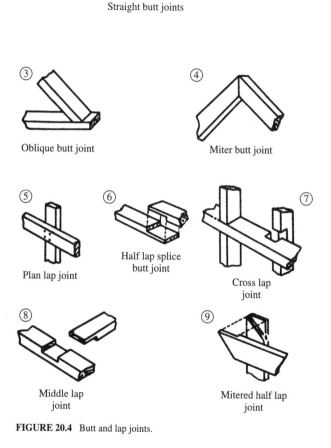

Straight butt joints

Oblique butt joint

Miter butt joint

Plan lap joint

Half lap splice butt joint

Cross lap joint

Middle lap joint

Mitered half lap joint

FIGURE 20.4 Butt and lap joints.

(Figures 20.6 and 20.7). If you use air gear, make sure it is set to sink the nails in deeply enough. If you're still hammering nails in by hand, don't forget your nail set. If your painters have to go around and set a bunch of nails, you're going to have some irate painters on your hands.

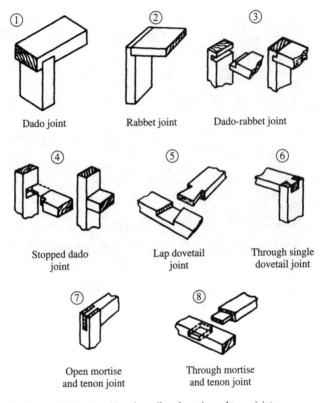

FIGURE 20.5   Dado, rabbet, dovetail, and mortise and tenon joints.

Finish	Characteristics
Polyurethane	Expensive Resists water Durable Scratches are difficult to hide
Varnishes	Less expensive than polyurethane Less durable than polyurethane Resists Water Scratches are difficult to hide
Penetrating sealers	Provide a low-gloss sheen Durable Scratches touch up easily

FIGURE 20.6   Wood finishes.

- Stain trim prior to painting walls and ceilings.
- A bristle brush is an excellent tool to use when applying stain, wipe off excess stain with a clean rag.
- Trim made of soft wood, such as pine, should be treated with a sealer prior to staining.
- When staining is complete, sand and dust the trim before applying polyurethane or varnish.

FIGURE 20.7   Tips for staining trim materials.

NOT MUCH TO SAY

There is really not much to say about interior trim that a professional doesn't already know. The key to interior trim is patience. Surprises rarely pop up when installing trim. Choices of standard trim are not numerous. The simplicity of installing trim doesn't allow for long discussions. So, rather than waste your time with long-winded, round-about talk pertaining to trim, let's move into the next chapter, where we are going to talk about cabinets and counters.

CHAPTER 21
DECKS AND PORCHES

Decks and porches are common add-ons for homes. A lot of homeowners desire a nice deck. Many people prefer a screened porch. Some property owners opt for a glass-enclosed porch. The construction of these projects is all very similar. There are, of course, differences in the projects, but the basics are quite similar.

As with most elements of construction, there are many acceptable ways to complete the creation of a deck or porch. Contractors all have their own opinions and ways of making a suitable addition. Here, we will go over the basic components and concepts used to construct both decks and porches.

FOOTINGS

Footings are one of the first steps in building a deck or a porch. We talked about footings in chapter 5, but let's go over a few of the details that pertain to decks and porches.

Most decks have pier foundations. This means that the footings are made as a series of independent concrete pads for support posts to sit on. Porches can be constructed on pier foundations, but they are often built on running, or continuous, footings. Pier foundations are, obviously, less expensive to build than continuous footings. You can simply dig a hole and fill it with concrete to create a pier footing, but a form tube, available at building suppliers, makes the pier footing more uniform and can reduce the risk of frost heaves in cold climates.

Footings of any type must be installed well below the local frost line. Check your local code requirements to determine what your local frost line is, and make sure the footing is at least 6 inches below that level. Deeper is better.

SUPPORTS

Supports for porches and decks can be the same or different. It is common to use pressure-treated wood posts to support decks. Many porches are held up by this same type of support (Figure 21.1). However, a concrete, block, or brick-and-block foundation is also common for a porch. A full foundation makes for a warmer floor, which can be important in the case of a heated porch. However, this, of course, is not needed for a deck. Screened porches are often built on piers, but a full foundation might be used to appearances.

The size of support timbers varies with construction details and local codes. A deck or porch might be supported with 4-×-4 posts or 6-×-6 posts (a better choice when affordable).

It's possible for posts to simply sit on pier footings, but it is much wiser to use a post anchor to secure the post. Anchors of this type are pressed into the pier while the concrete is still soft. Once the concrete sets up, the anchor is solid

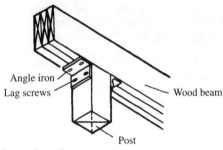

A. Connection to beam

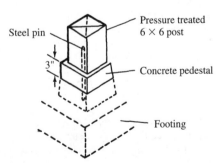

B. Base support

FIGURE 21.1 A typical deck foundation support detail.

Beam size	Maximum distance between support post
4-×-6	6 feet
4-×-8	8 feet
4-×-10	10 feet
4-×-12	12 feet

FIGURE 21.2 Common beam spans for decks.

Beam size	Maximum span
2-×-6	8 feet
2-×-8	10 feet
2-×-10	13 feet

FIGURE 21.3 Common joist spans for decks (16 inch on center).

and will accept the support posts, which are attached to the anchor with nails or screws, therefore, reducing the risk of post movement in a lateral direction.

Once support posts or foundation walls are in place, a sill plate and a band board are needed. The sill sits on the foundation and provides a place for floor joists to sit. A band board stands vertically and allows a place for the joists to be nailed to. Ideally, metal joist hangers should be used to support any joist that does not have a sill under it, such as where joists meet at a center beam. Some contractors use a wood ledger, such as a 2-×-4 on beams to support the bottoms of joists.

The framing that attaches a deck or porch floor structure to a home must be flashed and secured. The flashing will go up under existing siding and come down over the band or ledger board that is secured to the home with lag bolts. The ledger or band board will have a thickness equal to that of the floor joists.

Floor joists will generally sit on beams (Figure 21.2) that are in the middle of the framing detail. Each end of a joist will be secured to a band board, and metal joist hangers are recommended. Blocking, or bridging, between the joists will keep them from twisting. There are times when bracing might be installed between support posts and beams to add more strength to the construction, but this is not common in most applications (Figures 21.3 through 21.5).

Once the framing detail is complete, the flooring can be installed. For a deck, this will probably be done with planks. A porch floor is likely to be covered with a subfloor of 4-×-8 sheets of material. Decks will require the installation of railings and perhaps steps. Porches require full walls, a roof structure, siding (or screen) door, and a roof. This, of course, makes porches more complicated and time consuming.

Beam size	Maximum span
2-×-6	6 feet
2-×-8	8 feet
2-×-10	10 feet

FIGURE 21.4 Common joist spans for decks (24 inch on center).

Beam size	Maximum span
2-×-6	5 feet
2-×-8	7 feet
2-×-10	8 feet

FIGURE 21.5 Common joist spans for decks (32 inch on center).

DECK MATERIALS

Deck materials can cover a wide range. You can use redwood to create a notable deck, but this will be very expensive. Pressure-treated lumber is the most common deck material, but it has a greenish tint and doesn't hold paint well. The wood you choose must be durable when exposed to exterior conditions. Cedar is a good option if you don't want to use pressure-treated materials. Nails should be galvanized or stainless steel, otherwise, rust streaks are likely to taint your work of art.

PORCH MATERIALS

Enclosed porches that will be heated and used as typical living space can be framed with standard building materials. The construction of a common enclosed porch is no different from building any other part of a home. Sunrooms require a lot of glass, and this can be accomplished in a number of ways. Sliding-glass doors are an economical way to create walls for a sunroom. However, sliding-glass doors allow ventilation through only one-half of the glass space. Casement windows allow full ventilation and might be a better choice. Sliding-glass doors force the use of floor-mounted electrical outlets. When a short wall is used in conjunction with windows, wall outlets can be installed.

Sunrooms are often more user-friendly around children when a typical wall is extended upwards to accept casement windows. This keeps the kids away from glass at the floor level. Furniture placement is also easier when half-walls and windows are used, instead of sliding-glass doors.

Screened porches are commonly constructed of pressure-treated lumber. Screen is installed to protect the space from insects. The screen is secured with the use of lattice strips. Roof construction for any porch is about the same as any other roof job. The rafters and roof coverings are the same, but ceiling joists will not normally be used in a screened porch.

Offering your service as a contractor who builds porches and decks can add considerably to your annual income. Both porches and decks are very popular, so there is plenty of opportunity to gain additional work when you move into this area of home improvements.

CHAPTER 22
FIREPLACES AND FLUES

Fireplaces, wood stoves, flues, and chimneys can be part of any job. Some people want a wood stove installed to use as a back-up heat source. Other people want them to use as an alternative heat source. Fireplaces are desirable in the eyes of many people. With either a fireplace or a wood stove, a chimney of some type is required. The chimney made be made of metal pipe or masonry materials. Fireplaces can be prefab metal units that don't require a lot of alteration in existing construction, or they can be masonry monsters that mandate a footing and substantial changes to existing construction conditions. Some contractors are not familiar with what is involved in the installation of a fireplace or wood stove. For this reason, we are going to go over the basics. Before we do, however, I want to stress that you should consult with local authorities with regard to current local code and safety requirements before installing any type of fireplace or stove.

CHIMNEYS

Chimneys have been made with little more than sticks and mud. This type of construction is no longer acceptable. Today's code requirements for chimneys and flues are much more restrictive, and with good reason. Unlined brick chimneys often lead to unwanted chimney fires. Mud and sticks are not a prime choice in terms of chimney materials. However, even after throwing out some of the more primitive methods for building a chimney, we are still left with options. Do you know what they are?

Masonry

Masonry chimneys are looked upon as being ideal (Figure 22.1). They are expensive, and they need to be cleaned periodically, but they are generally

Minimum net interior area (square inches)	Nominal dimensions (inches)	Outside dimensions (inches)
15	4 × 8	3.5 × 7.5
20	4 × 12	3.5 × 11.5
27	4 × 16	3.5 × 15.5
35	8 × 8	7.5 × 7.5
57	8 × 12	7.5 × 11.5
74	8 × 16	7.5 × 15.5
87	12 × 12	11.5 × 11.5

FIGURE 22.1 Modular clay flue lining sizes.

1. Fireplaces: Flue should contain a minimum of 50 square inches and be no less than one-half the size of the fireplace opening.
2. Boilers and furnaces: Flue should contain a minimum of 70 square inches.
3. Room heaters and typical stoves: Flue should contain a minimum of 40 square inches.
4. Small stoves and heaters: Flue should contain a minimum of 28 square inches.

FIGURE 22.2 Recommended flue sizes for residential applications.

considered to be the best type of chimney available. I don't disagree with this view completely, but neither do I feel that a masonry chimney is always best.

When a masonry chimney is built, it must have a solid platform to rest on. This typically entails the use of a footing. If the chimney is to be installed on the outside of a home, substantial work must be done to cut it into the siding. Exterior masonry chimneys are usually covered in brick for appearance purposes, and this runs their cost up. If I were installing a wood stove, I probably would not opt for a masonry chimney. However, if a masonry fireplace is being built, it will be served by a masonry chimney. So, how involved is it for a general contractor to undertake the installation of a masonry chimney for an existing house or a new home?

The first choice someone has to make is whether the chimney will be installed within the home or on an exterior wall (Figure 22.2). Interior chimneys can be enclosed with standard building materials, and this eliminates the need for a brick exterior, with the exception of where the chimney exits a roof. When the brick is avoided, there is a substantial savings in cost.

Before a chimney can be installed inside a home, a proper space for it must be found. Local codes will set requirements for clear space around the chimney. When planning an interior chimney location, you must take the clearance requirements, the size of the chimney, and the concealment framing all into account. The amount of room needed for all this can be quite a bit. So much, in fact, that an interior chimney of this type might not be practical.

If a chimney is to be installed along an exterior wall, outside of a home, a footing and concrete pad will be needed for support. Existing siding on the home will have to be cut so that the chimney can be built along the sheathing of the home. Then flashing will have to be installed along the length of the chimney. All of this runs the expense of such a job up. Is your customer willing to pay thousands of dollars for a chimney? It could easily cost that much to build a masonry chimney. Is there an alternative? For wood stoves and pre-fab fireplaces there is.

Metal Chimneys

Metal chimneys are far from cheap, but they are much less expensive than their masonry cousins. Double- and triple-wall metal chimneys can be installed with a minimum amount of clearance. This allows them to be installed in smaller spaces than a masonry unit could be. Almost anyone can install a metal chimney, so the high cost of a mason is not necessary. An average remodeler can easily install a metal chimney, from start to finish, in less than a day. This is much faster than the time it would take to have a masonry chimney built, and there are other advantages to metal chimneys.

Unless a customer wants a brick chimney for status or appearance, there is little reason to use a masonry chimney for anything other than a masonry fireplace. Metal chimneys can be installed inside or outside of homes. When installed outside, the chimney pipe can be enclosed with framing and siding. Inside, the pipe can be hidden with framing and drywall. Framing and siding can be installed around the chimney where it exits a roof (Figure 22.3).

The relative low cost of a metal chimney makes it not only a viable option, but an affordable one. Because metal chimneys have a very smooth finish on the inside, they are not as prone to catch and hold creosote. This is not to say that they shouldn't be cleaned, but they might be less dangerous than a masonry chimney that has rough surfaces along its channel.

Few general contractors or remodelers are accomplished masons. If you have a customer who wants a masonry chimney, you should have an experienced mason visit the job site with you before making any commitments. The mason will be able to point out alterations that will be needed on the existing construction. This helps to protect you, as well as making your estimate more accurate.

When a metal chimney is suitable, you might have your own crews take care of the installation. Suppliers of metal chimneys usually stock a variety of

---

A chimney must extend at least 3 feet above a roof line or at least 2 feet above the highest point of any roof within 10 feet.

FIGURE 22.3 Chimney roof clearances.

kits and accessories to make installations safer and easier. For example, you can get a through-the-wall kit or a through-the ceiling kit. Each kit will contain special fittings that are designed to provide proper clearance and protection when a chimney pipe penetrates a wall or ceiling.

Let's say, for example, that you have a customer who wants a wood stove installed in a basement where you are doing a conversion job. The house has a basement and one level of living space. There is a standard gable attic over the living space. Your customer wants the chimney to run up through the house and has agreed to forfeit a small section of a room to house the chimney. What will you need to make this installation?

The first thing you will need are three collars that will be used where the chimney comes through the floor, ceiling, and roof. The collar will mount between joists, although you might have to cut out and head off enough space to accommodate it. The first collar will be installed in the basement. The exposed side of the collar will accept standard stove pipe. The upper side of the collar will accept the metal chimney pipe. This provides support for the chimney and complies with clearance requirements.

A section of chimney pipe is attached to the first collar, and subsequent sections are installed, one on top of the other until the ceiling area is reached. At this point, another collar is needed to make the penetration through the ceiling and into the attic. Chimney pipe is installed on both sides of the collar and is extended up through the attic. A hole is cut in the roof, and a third collar guides the chimney pipe through the roof. Once the pipe is out of the house, it is extended to meet local code requirements. A spark arrestor will normally be installed and so will a chimney cap. Other supports and accessories might be needed, depending upon individual conditions.

Once the chimney is installed and inspected, it can be concealed in a framed chase. Some clearance from combustible materials will be required by local regulations, but the distance will be minimal. This type of installation is fast, easy, safe, and affordable. It is hard to beat.

If the customer had wanted the chimney to extend up the exterior of the home, a through-the-wall kit could have been used. A special collar is used where the chimney penetrates the exterior wall, and then, a wall bracket provides support for the vertical chimney on the outside wall. The pipe is run up the side of the house, using special brackets to hold it in place. When the piping is complete, a wood frame can be built around the chimney, leaving required clearance, and covered with siding. Either of these types of installation are much simpler than what would be required for a masonry chimney.

FIREPLACES

There are three basic types of fireplaces that you might be asked to install. The most common, and most expensive, is a full masonry fireplace. A second type of fireplace is a prefab unit that is designed to be built into a wall. The third type of fireplace is a free-standing unit. There are pros and cons to each of these types of fireplaces, so let's discuss them.

Masonry Fireplaces

Masonry fireplaces are typically considered to be the most desirable. They can be made to fit almost any location nicely, and they are extremely durable. The biggest drawbacks to masonry fireplaces are their cost to install and the work involved with the installation (Figures 22.4 and 22.5). Adding a masonry fireplace to an existing house is no small job. It is a major undertaking that will consume many days and a lot of dollars.

The consideration many homeowners look at when assessing fireplaces is the affect they will have on the appraised value of their homes. From my discussions with real estate appraisers, masonry fireplaces do very well when it comes time for an appraisal. While it is unlikely that excess equity will result from adding a masonry fireplace, it is likely that most of the cost will be returned in appraised value. Considering the fact that a masonry fireplace is likely to cost a minimum of $4000, this is comforting to know.

Built-in Fireplaces

Built-in fireplaces cost a fraction of what a masonry fireplace does. Prefab, metal fireplaces can be installed in any room without excessive alteration. Installation

Brick type	Dimensions (inches)
9" straight	$9 \times 4\frac{1}{2} \times 2\frac{1}{2}$
9" small	$9 \times 3\frac{1}{2} \times 2\frac{1}{2}$
Split brick	$9 \times 4\frac{1}{2} \times 1\frac{1}{4}$
2" brick	$9 \times 4\frac{1}{2} \times 2$
Soap	$9 \times 2\frac{1}{4} \times 2\frac{1}{2}$
Checker	$9 \times 2\frac{3}{4} \times 2\frac{1}{4}$

FIGURE 22.4 Firebrick sizes.

Brick type	Laid flat	Laid on edge
9" straight	6.5	3.5
9" small	6.5	4.5
Split brick	13	3.5
2" brick	8	3.5
Soap	6.5	7
Checker	6	6

FIGURE 22.5 Number of firebricks needed per square foot of coverage.

is simple. The unit is set in place, metal chimney pipe is run, and a wall is framed up around the fireplace and chimney. These working fireplaces can add a touch of romance to a master bedroom or warmth to a family room.

What is the major drawback to a metal fireplace? Well, it isn't cost, because they are cheap in terms of fireplaces. I've had both masonry and metal fireplaces in homes where I have lived. Without question, I have preferred masonry fireplaces. My personal experience as a user of a metal fireplace is that the firebox is too small. The space is typically short and narrow, which restricts log length and burning time between trips to the wood pile. Other than this one complaint, I don't know of any other serious drawback.

Free-standing Fireplaces

Free-standing fireplaces were very popular for awhile, but the infatuation with them seems to have waned. These are units that are intended to set out in the floor of a room. A chimney pipe extends of the top of the unit and is usually left out in open view. These fireplaces are inexpensive, in relative terms, and they are easy to install. Some are very attractive and quite functional. The free-standing aspect makes these fireplaces more like a wood stove than a fireplace, in terms of heating capabilities.

One potentially dangerous drawback to a free-standing unit is the risk of someone getting burned. This is especially true when small children will be found in the vicinity. The relative low cost of free-standing and prefab fireplaces make up for the fact that neither of these units do great in terms of appraised value.

WOOD STOVES

Wood stoves don't require as much work to install as most fireplaces do. A chimney is built and the wood stove is set in place. A stove pipe connects the stove to the chimney collar, and the job is done. There are, of course, a lot of wood stoves to choose from on the market. Making a decision on which stove to buy might very well be the toughest part of the job. Fortunately, your customers won't expect you to hold their hand while they shop for a stove. As long as you are aware of local code requirements and the methods for installing a safe, acceptable chimney, you are in the clear on wood stoves.

If your customer is planning to have you install a stove, you might suggest a brick hearth and perhaps even a brick heat shield (Figures 22.6 and 22.7) for the wall behind or on either side of the stove. When a suitable heat shield is used, a stove can be safely set closer to a wall. This conserves floor space for your customer. The heat shield might be made of metal, brick, tile, or some other approved material.

- The back of a stove pipe should be at least 18 inches from any combustible material.
- The back of a stove should be at least 36 inches from any combustible material.
- The side of a stove should be at least 36 inches from any combustible material.
- A distance of at least 18 inches should be maintained with some sort of fire-resistant base for the stove to sit on.

FIGURE 22.6 Clearances for wood stoves without heat shields.

- The back of a stove pipe should be at least 9 inches from any combustible material.
- The back of a stove should be at least 18 inches from any combustible material.
- The side of a stove should be at least 36 inches from any combustible material.
- A distance of at least 18 inches should be maintained with some sort of fire-resistant base for the stove to sit on.

FIGURE 22.7 Clearances for wood stoves with heat shields in rear.

OTHER CONSIDERATIONS

Other considerations might come up when talking with customers about wood stoves and fireplaces. For example, your customer might want gas piping run for a gas log. In most jurisdictions, a licensed gas fitter will be required for such an installation. Many licensed plumbers are also licensed gas fitters. If the customer is seeking a sensible source of heat, your conversation might change to gas- or oil-fired wall heaters. Some gas-fired heaters are rated to work safely without any venting to outside air. Most wall units do, however, require a direct-vent system, which is easy to install.

It is difficult to stay up to speed on all the many options available to homeowners when it comes to remodeling. If you sat in your office all day, every day, reviewing trade publications, brochures, and other available data, you might be able to absorb most of what's hot in the market. However, few remodelers have time to sit around ferreting out new products and their specifications. You should strive to stay informed, but don't be afraid to ask a customer to allow you to research a subject. If you are asked a question that you are unable to answer with authority, don't bluff the homeowner. Request some time to look into the matter and then present your findings at a later date. It is better to say nothing than to say something that is not correct.

CHAPTER 23
CABINETS
AND COUNTERS

If you're remodeling a kitchen, you are almost certain to work with cabinets and counters. Bathroom remodeling can also put you into contact with these elements. If you're a seasoned remodeler, you already know how much a set of cabinets can cost. You are probably also aware of the broad spectrum of prices affixed to cabinets. Counters can also be plenty expensive. If you've ever miscut a sink hole in a new countertop, you probably know all too well how much a counter can cost. When you consider that a set of kitchen cabinets can cost less than $2000 or more than $10,000, it is easy to see why the subject is worthy of your attention.

Builders also know, full well, how much cabinets can cost. When a builder must make a financial arrangement for a cabinet allowance in a bid, the range of costs can be extreme. Cabinets and counters are, without a doubt, major expenses in building and remodeling.

Kitchen remodeling is certainly one of the most popular forms of remodeling for homeowners. Statistics show that kitchen remodeling is one of the safest home improvements a homeowner can invest in. With such a current interest in kitchen remodeling, there has been a surge in the cabinet industry. Cabinet manufacturers are offering more designs and styles than ever before. There are do-it-yourself videos available to show homeowners how to reface their existing cabinets. Major retailers are co-venturing with remodelers to offer replacement doors and drawer fronts to be used in giving old cabinets facelifts. When you sit down with a potential customer for a kitchen job, you might have to explain the pros and cons of refacing, replacement doors, and new cabinets. Are you prepared to do this? Let's find out.

AN AVERAGE HOMEOWNER

Assume that I'm an average homeowner. I've asked you over to discuss the options for improving the look of my kitchen. As we sit at the dining room table, I begin asking you questions. I'm expecting some legitimate answers, so let's see how you do.

NEW DOORS

I've heard about ways to give my existing cabinets a new look by installing new doors and drawer fronts, what can you tell me about this procedure? How will I make the rest of my cabinet surfaces match the new doors and draw fronts? Where will the new hardware come from? Will I have a choice in door pulls and hinges? How much money will I save by having my old cabinets updated instead of replaced?

How are you doing with your answers? The replacement of doors and drawer fronts is a common procedure. This work goes quickly. It can be completed in a day or less. As for the rest of the cabinet surfaces, they can be covered with a veneer or an adhesive paper that will match the new doors and drawer fronts. Hardware can come from any number of sources. It is available from standard suppliers of building supplies, as well as specialty suppliers. The amount of money saved by refacing instead of replacing can be quite substantial. An exact amount is difficult to arrive at on a generic basis, but it could easily amount to thousands of dollars. Now, let's get back to our make-believe estimate interview.

Can you tell me the pros and cons of a wood veneer over a wood-look adhesive paper? How long will a refacing job last? Explain to me the advantages of replacing my old cabinets. How long will it take to get new cabinets? Would you recommend store-bought cabinets or custom-made cabinets? Let's review these questions and some potential answers.

If money is of a major concern, wood-grain adhesive paper can be used to give a new appearance to old cabinets. This is the cheap way out, and it does come with some potential problems. The paper can be tricky to install. Keeping wrinkles and bubbles out of the paper is not always easy. Because the covering is paper, it can be cut or torn. If this happens, the whole facelift can be ruined. Sometimes a patch can be made, but more often than not, an entire piece of the paper has to be replaced.

Wood veneer is a better option than adhesive paper when putting a false front on kitchen cabinets. The veneer is usually applied with a contact cement. Because the veneer is made of wood, it looks real. The veneer won't tear or rip, and it passes the touch test. In my opinion, wood veneer is the only way to go when refacing cabinets.

A quality refacing job can last years. It might not be as durable as a new set of cabinets, but for the cost, it is a good value. There are some advantages to doing a complete replacement, instead of a refacing job. When you are forced to work with existing cabinets, you are limited in your options. Tearing out old cabinets and replacing them with new ones will provide an opportunity for the creative use of all types of cabinets. Turn-table cabinets can be put into the system. Cabinets that house recycling bins can be installed. A myriad of possibilities exist when you start from scratch.

Timing can be critical when dealing with kitchen cabinets. It can take months to have a set of custom cabinets made and delivered. Stock cabinets, on the other hand, might be available for immediate pick-up. Most production cabinets can be ordered and delivered in less than three weeks. All of this comes into play when making a buying decision.

With the many alternatives available in production cabinets, there is little reason to pay extra and wait longer for custom cabinets. The quality of production cabinets has reached a level where it can be very difficult to tell a stock cabinet from a custom cabinet. When you weigh out the price and the delivery time of each type of cabinet, production cabinets will win the race every time. There are certainly occasions when custom cabinets are in order. If a person wants a unique cabinet arrangement, custom units will be the only way to go.

COUNTERS

Helping your customers make decisions on counters can be almost as difficult as helping them choose cabinets. If nothing else, the multitude of colors and patterns will keep your customers confused for a good while. Will you recommend a square-edge counter or a round-edge unit? What type of backsplash will you suggest? Are you going to offer to make up a counter onsite, or will you have your customer order a top from one of your suppliers? Is a laminate top the only type you plan to offer your customers? Have you considered a tile counter? What about a marble-type counter? Do you think your customer will prefer a slick finish or a pebble finish? Did you know there were this many questions that could come at you so quickly just on the subject of counters? Customers you sit down with might hit you with a lot more questions, and you should be prepared to answer them.

Many people don't give much thought to their counters. They go to a showroom, pick out a color and a pattern, and tell their contractor to order it. It is usually a safe bet that a customer will order a butcher-block design in neutral browns and tans. I know this is the type of laminate top that I get the most demand for.

If you want to take the path of least resistance, you can do as your customer says and order the top. However, if you want to make a good impression, and possibly a little extra money, you can educate your buyer in the other counters that are available.

Many homeowners never stop to think about any type of top other than what they have been used to. If the people have lived in apartments and tract housing, a basic laminate top is what they will assume is standard procedure. Can you imagine how these customers might respond if you show them pictures of nice tile counters? If you will be using the job as a reference, it is in your best interest to make the work as nice as possible, and this includes making it a little out of the ordinary. I'm not suggesting that you twist a customer's arm to get a flashy counter installed, but it never hurts to apprise people of their options.

Let's assume your customer is dead set on a plain laminate top. Can you sell them on the idea of installing a tile backsplash between the top and the bottom of their wall cabinets? If you can, there will be a few extra bucks in the job for you, and you will have a better grade of work to use as a reference. The customer will get an easy-to-clean wall of tile, and you will be an above-average remodeler in regards to your base grade work.

Do you ever build your own counters (Figure 23.1)? If you don't, you should consider it. At the very least, you should find a subcontractor who can do custom, onsite countertops. If you stay in remodeling long enough, you are sure to have an occasion when a site-built top will be of great interest to a customer. I can give you an example of this from one of my recent jobs.

I was putting together a new kitchen package a few months ago. The job involved a corner sink. With the double-bowl corner sink, a stock laminate counter was not a great idea. The seams for the top would meet under the

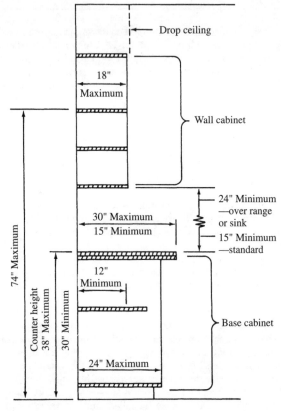

FIGURE 23.1 Common kitchen cabinet dimensions.

drains of each sink bowl. Not wanting seams where water infiltration might be likely, it was necessary to look for an alternative. In this particular job, a post-form top would have cost me about $300. However, because this type of stock top was not desirable, due to the location of seams, I investigated other options. One of these alternatives was a more custom type of top. This counter would be seamed away from the sink, but the price for the top would have been a little more than $600. None of these prices included installation.

After weighing my options on prefab counters, I looked into having the top built onsite. Building the counter right in the kitchen would allow me to avoid seams under the sink, and it would speed up the job. Instead of waiting two weeks for an ordered top, I could have one of my subcontractors build the counter in less than a day. The price I was quoted, for labor and material was less than $450. This option gave me a custom-built counter, which could be installed immediately, for about $150 less than a prefab top that didn't include installation. I chose to have the top built onsite, and it turned out great.

If I had wanted to, I could have easily charged $600 for the counter and a few hundred more for installation. This price would have been in keeping with the cost of a prefab top, and the job would have still been finished quicker. I would have also been about $350 richer. Not being greedy, I didn't inflate the cost of this work. However, I could have pocketed the extra cash without really taking advantage of the customer. The homeowner would have gotten a better counter in less time for the same money as a prefab unit. This example alone proves the value of being prepared to do onsite work with counters.

BATHROOMS

When you are remodeling bathrooms, you will probably be installing vanity cabinets and tops. This, of course, is no big deal. Almost anyone can install a vanity base and a molded lavatory top. In many cases, space limitations restrict creativity with bathroom cabinets. However, when there is adequate room to work with, a vanity can become a work of art.

Most contractors install production cabinets in bathrooms. I've had custom cabinets made for vanity bases from time to time, but a vast majority of my work has been done with stock cabinets. The hardest part about this type of work is helping the customer decide on a particular type or design. When there is enough room to work with, the options for a vanity base are considerable.

If you only have a 30-inch space to work with, you're not going to be able to do much in the way of exotic cabinetry. This type of small space will basically limit you to a base cabinet with one door and no drawers. There are a few cabinets available in this size with drawers, but they are not common. The larger the space you have to work with, the more you have to offer your customers.

If the space available in a bathroom for a vanity is long enough, you can suggest a double-bowl vanity. Customers can choose base cabinets that have doors under the lavatory bowl and drawers on one side or even on both sides. A fairly standard layout is a cabinet with a door under the lavatory bowl and drawers on both sides. This provides his-and-her drawer space.

One design that has been very popular with my customers is a base cabinet with an offset top. The cabinet houses a lavatory and the top extends for some distance to create a make-up counter. Add some strip lights, a big mirror, and a seating arrangement, and the homeowner has a fantastic place to get ready for work or play. This type of dressing area is very popular in the areas where I've worked.

Vanities can be focal points of bathrooms. The wood, the doors, the shape, and the size of a vanity can all work to create a haven, rather than a simple bathroom. When vanities are used, counters are needed. These can be cultured marble tops with integral lavatory bowls or laminated tops with drop-in or self-rimming lavatory bowls. Self-rimming bowls look better, are easier to clean around, and are not as prone to leaking. Cultured marble tops are usually chosen when a vanity is used. Sometimes a more expensive type of top is picked, and a few people prefer a laminate top. My experience has shown, however, that a cultured marble top, with an integral lavatory bowl, is a big favorite over other options.

CONSTRUCTION FEATURES

Customers might want you to point out construction features as they apply to cabinets and counters. This is not very difficult if you know your product line. What are some of the features that you recommend customers to look for in a quality cabinet? Views on what makes a good cabinet can vary from person to person, but there are some benchmarks with which most contractors agree (Figure 23.2).

Type of cabinet	Features
Steel	Noisy Might rust Poor resale value
Hardwood	Sturdy Durable Easy to maintain Excellent resale value
Hardboard	Sturdy Durable Easy to maintain Good resale value
Particleboard	Sturdy Normally durable Easy to maintain Fair resale value

FIGURE 23.2   Kitchen cabinet features.

Type of cabinet	Price range
Steel	Typically inexpensive, but high-priced units exist
Hardwood	Typically moderately priced, but can be expensive
Hardboard	Typically moderately priced
Particleboard	Typically low in price, but can reach into moderate range

FIGURE 23.3 Kitchen cabinet price ranges.

Most contractors agree that good cabinets are enclosed fully, meaning that they have backs in them. Not all cabinets do. Cabinets without backs are not as sturdy as those that do have backs. How much of the cabinet is made of wood? The more wood that a cabinet contains, the higher it is usually thought of. Were butt joints used in the construction of the cabinet? Dovetail and mortise joints are recognized to be of a higher quality than butt joints. Do the drawers of a cabinet open and close with ease? Drawers should be set into place on smooth glides. If a drawer is jerky to open or close, the quality of the cabinet is probably low. This type of problem could result from a slopping installation, but it is more often the sign of a cheap cabinet. Are the shelves in the cabinet adjustable? They should be. How many adjustment options are there, and how difficult is it to move shelves around? Good cabinets offer a wide variety of possible shelf heights, and the movement of shelves should be simple.

I believe that customers should see and try cabinets before they buy them. If you are selling stock cabinets, your customers should be able to go to a showroom and compare various styles and types. When custom cabinets are going to be made, the customer can't see and touch the exact cabinets that will be used until they are made. However, a cabinet maker should have samples of the types of cabinets offered for sale. These samples can provide a glimpse of what to expect in custom cabinets being ordered.

The cost of kitchen cabinets can be extremely high (Figure 23.3). With so much money on the line, customers owe it to themselves to try before they buy. In other words, they should go out and look around at all types of cabinets. Particular attention should be paid to construction features. A cabinet that has terrific eye-appeal might be a piece of junk. Getting a cabinet with rotating shelves that don't turn smoothly can be disappointing. Wall cabinets with doors that won't stay shut are a nuisance. Drawers that stick and scrape are no good. You should take your customers to showrooms and let them play with sample cabinets. Encourage them to open and close doors, spin turntables, use drawers, and evaluate all the various types of cabinets.

While you are at the showroom with your customers, you should discuss door styles and hardware. Will the doors have raised panels? Do leaded-glass doors fit your customer's budget and design? Will the customer want finger grooves or door and drawer pulls? Answer as many questions as you can in the showroom. There is no better place to work through confusion with cabinets and counters than in a showroom.

BALANCE

It can be difficult to balance a budget and desire at the same time. When I figure a job that will involve new kitchen cabinets, I plug in an allowance for the cost of the cabinets. For example, my proposal will state that my price is based on a cabinet allowance of $2800 (at my wholesale cost). The customer can spend more or less. However, because I have no way to know what taste in cabinets the customer has, I have to use a random figure. When I pick an allowance figure, I base it on my gut feeling. Some people spend less than $1500 on cabinets. Others spend $7000, or more. On average, I've found the wholesale cost of cabinets for most of my jobs to range between $2500 and $3500.

When you take your customers shopping for cabinets, you are very likely to see them go through a bit of agony. They know they have a budget to work with, but they have trouble deciding what to do. On one hand, they want to choose something that costs less than their allocated sum. This will save them money on the overall cost of the job. However, then they will have a higher-priced cabinet style that they like better. It might very well be within their budget, but pushing it. Then, there will be the cabinets that they really love. These units, of course, are way above the budgeted price. So, what is the customer to do? You might be asked to help settle the dilemma.

I recently had a situation just like the one I've described. My wife and I built a new house. During the cabinet stage, we went to the showroom to pick and choose. We knew from the start what brand of cabinet we wanted. After building 60 homes a year and running a very active remodeling business for over a decade, we had no doubt about the cabinet manufacturer that we trusted most. However, we were faced with the money crunch. With as much experience as we both have in the construction and remodeling field, Kimberley and I still got caught in the pricing web.

We had set a budget of $2500 for our kitchen cabinets. For the first time in any of the houses we have built for personal use, we wanted white cabinets. When we visited the showroom, we were not surprised to see a broad selection of styles and options. We're used to this from being in the business. However, we were like many other consumers who are trying to decide on which cabinets to use: confused!

All of the cabinet styles we were looking at were of a good quality. The least expensive design was all white. The cabinets had flat doors and finger grooves. This particular set of cabinets would have cost us about $1800. We contemplated the opportunity to save $700 on our allowance. I felt, however, that the cabinets were too clinical looking. They gave me the impression of being in a medical facility. We moved on to the next pricing level.

The second set of cabinets we examined would cost about $2300. This was still less than our budgeted amount, and the cabinets looked more residential than the first group did. They still didn't feel quite right. They were livable, and we were both willing to accept them, but neither of us were thrilled with them. This led us to a third level.

When we reached a higher price range, about $3300, we got to cabinets that were white with oak trim. The oak strips around the finger grooves were a nice touch, and they wouldn't show dirty fingerprints the way a stark white cabinet would. Kimberley and I both really liked the oak accents. After considerable discussion, we decided to go with the oak. All of a sudden, we were $800 over budget, but how many houses do most people build for themselves? We've lived with the cabinets now for about four months, and we still love them.

I've told you this story to put into perspective what your customers might go through at a showroom. With all the years of experience I have in remodeling and construction, I still struggled with a cabinet selection. If someone like myself can have this type of problem, you can image how an average homeowner might feel. As a professional, you should be available to assist your customers in making their decisions. The people selling cabinets are going to stress sales features. It is up to you to show your customers all angles of a cabinet selection. If you take the time to explain construction features, cost factors, market appeal, and other elements of cabinets, your customers will make decisions based on facts, not just sales hype.

LOWBALL PRICES

How many times have you seen lowball prices on cabinets in sales flyers? Almost any building supply center that caters to homeowners will carry some type of low-priced cabinets that can be advertised to pull people in. Many times the prices are not the bargain that they seem to be. It is not unusual for the rock-bottom prices of cabinets to exclude the finishing of the cabinets. I've had customers come to me with flyers and express a desire for the low-priced cabinets. It is true that unfinished cabinets can sell for half of what a standard production cabinet that has a permanent finish does, but this doesn't make the cheap cabinets a good deal. Even if the quality of the cabinets is not lacking, the cost and trouble of finishing them can outweigh any price savings.

I have a policy that stipulates the use of prefinished cabinets. My company will not assume any responsibility for finishing unfinished cabinets. We will install them if a customer insists, but we will not apply the finish. Finishing cabinets to a uniform look is very difficult and time consuming. It is not an area of work which I perceive as being profitable for most remodelers. Unless you have resources beyond those of most contractors, you will probably be better off in avoiding unfinished cabinets.

INSTALLATION PROBLEMS

Any time you are working with old houses, you are likely to have some installation problems with cabinets and counters. Floors aren't level, and walls aren't plumb. These conditions can make installing cabinets and counters very difficult indeed. The problems can be overcome if you catch them early.

However, when you wait until the cabinets are going in to discover the trouble, you've got a more serious problem.

Anyone who has been a remodeler for long knows that most houses are not perfect. In fact, they are usually far from it. If conditions are not right when cabinets are installed, problems will pop up. Doors might not stay shut. Drawers and slide-out accessories might not work properly. Gaps might exist along the bottoms of base cabinets. The space between a backsplash and a wall can be large enough to drop a pencil through. Face joints along cabinets might not fit up tightly. All of this can come together to give any remodeler who hasn't planned for the problems a splitting headache.

When you do an estimate for replacement cabinets, take a good level with you. Check the existing walls and floor. Even if you are going to gut the room out, check the existing conditions. Furring out walls and leveling floors takes time. If you don't budget this time into your job cost, you're going to lose money. When your level shows that problems with existing construction exist, you should show the trouble to your potential customer. They might not like finding out that additional work is going to be needed to bring existing conditions up to a satisfactory level, but they should appreciate finding out about the added expense before the job is underway.

Setting and hanging cabinets is not difficult work. It's not even very technical. However, when a room is not square, a floor is not level, or walls are out of plumb, the job can become tedious, to say the least. Existing walls can be furred out to make them suitable for cabinets. A floor that isn't level can be corrected with underlayment and, possibly, some filler compound. None of this work is a big deal if it is done during the rough-framing stage of a job. It can, however, be extremely difficult once the job is nearly finished and it's time to set cabinets.

Very few homeowners are going to appreciate having a wide piece of trim installed along the top of their backsplash to hide a huge gap. A small bead of caulking will be the most that they are expecting. If you have to put trim along the backsplash, someone messed up in their planning. It's standard procedure to shim up base cabinets, but the end result should not leave a noticeable gap along the floor. Yet, I've seen kitchen after kitchen where the base cabinets are jacked up to a point where shoe mold is needed to hide the imperfections. This might or might not be acceptable to a homeowner, but having such a gap shouldn't be necessary. If the floor is worked with prior to installing finish floor covering, there should not be any need for decorative trim to hide gaps.

The key to a smooth cabinet installation is planning. If you prepare a kitchen properly, installing cabinets will be simple and fast. Should you neglect to correct framing problems early on, finishing a job to the satisfaction of a customer can be all but impossible. Spend some extra time up front to avoid conflicts near the end of the job.

Most of our discussion has revolved around kitchen cabinets. This is because they are more numerous than bathroom cabinets. However, the same rules apply for vanities. You need to check closely to see that a new cabinet can be installed properly. You must also allow for the vanity top. This is something a lot of contractors forget to do. They think in terms of a 36-inch cabinet and fail to remember that the top will be 37 inches in width. This extra inch can be enough to cause you to pull some hair out. When you are measuring for cabinets, don't forget to allow for the countertops that will go on them.

DAMAGE

A lot of damage can occur near the end of a job, and some of it is likely to be with finished floors and cabinet installations. Using a screw that is too long can result in your having to buy a new cabinet out of your intended profits. If the screw punches through and ruins a cabinet, you are going to be responsible for the damage. Sliding base cabinets into place on a finished floor can cause other damage. If you cut, scrape, or tear a new floor, you will lose some more of your profit. Dropping screws and stepping on them is one way that many contractors inadvertently damage vinyl flooring. Some customers might accept a patch, but others will expect a complete floor replacement. This gets expensive, so you should make sure that you, and your crews, are careful not to harm new flooring.

Walls often suffer some damage during a cabinet and counter installation. This is to be expected, to some extent. Plan on having your painter do some touch-up work after an installation is complete. Ideally, this work should be done as soon as possible. The visual picture a homeowner gets when a new wall is scuffed or gouged is not conducive to referral business.

To create a complete list of potential problems to be encountered during a cabinet installation would take more space than we have. For example, a slip in cutting a sink hole into a cabinet can quickly mean a delay of weeks and lost money. A long screw coming up through a new counter is disastrous. Having a worker lean too hard against an open cabinet door can result in damage that is serious enough to warrant the replacement of a cabinet.

Almost all of the risks associated with damaging cabinets can be eliminated with good work habits and strong concentration. Take your time and do the job right the first time. You're getting paid for the initial installation; however, if your get careless, you might be paying for repairs and replacements.

CHAPTER 24
PROVISIONS FOR THE PHYSICALLY CHALLENGED

Physical challenges exist for all of us. However, some people are more challenged by daily life than others. Contractors are often asked to create or convert living space to meet the needs of individuals who require selective modifications in standard building practices. Some contractors are not familiar with the code requirements pertaining to barrier-free housing. This chapter will give you a good overview of what to look out for when building or remodeling living space to accommodate people who have special needs.

Physical challenges come in many forms. Some people must rely on crutches to get around. Walkers and wheelchairs are also required equipment for the mobility of some individuals. Age creeps up on everyone, and advancing years can restrict a person's flexibility and movement. There are, of course, many other conditions that can affect a person in a way that will require building modifications.

OUTSIDE A HOME

The conditions outside of a home can be critical to the use and enjoyment of a house for someone who has physical restrictions. For example, a person in a wheelchair can have a difficult time negotiating a driveway that is covered with crushed stone (Figure 24.1). In this case, a concrete or asphalt driving surface is much easier to roll a wheelchair on. Width is another consideration pertaining to the parking area. If a person has a van with a ramp that comes out to allow the use of a wheelchair, additional width might be a requirement in the parking area.

People who walk with the aid of crutches must be careful of slippery surfaces. Asphalt provides texture for the rubber tips of crutches to gain traction. If concrete is used as a driving surface for a person who is dependent on crutches, a rough texture, such as a broom-swept surface should be used on the finished driving and parking area. Essentially, you must evaluate the personal needs of each individual when deciding on the type of driving surface to install.

Maximum slope	1 in 12
Minimum clear width	3'0"
Minimum curb height	2"
Railing	2'8"
Maximum length	30'0"
Minimum width of passage	2'8"
Approximate length of chair	3'8"
Seat height	17" to 18"
Maximum comfortable reach	5'4"
Eye level	3'9" to 4'3"

FIGURE 24.1   Wheel chair ramps.

Garages

Garages are an excellent aid to people of all types, but they can be essential to people who have trouble walking or rolling through inclement weather. When building or modifying a garage, you must make allowances for additional width and perhaps a series of ramps for wheelchair use. Heavy-duty grab-bars might need to be installed to provide a stable access for individuals entering their homes. An automatic garage door opener, equipped with an automatic light, should be considered standard equipment.

Ramps

Ramps (Figure 24.2), made of concrete or wood, often replace stairs when a residence is designed for barrier-free living. A width of 4 feet is a good choice for a ramp. Even if people are not restricted to a wheelchair, ramps can make accessing a home much easier and less dangerous. Make sure that the ramp is built with a grade that conforms to local building codes, and equip the ramp with a slip-resistant surface.

Porches

Porches should be covered with a roof to protect people entering a building. The dimensions of a porch should be large enough to allow easy turning for a wheelchair, if a wheelchair will be used. Grab bars near an entrance door can help a person steady posture while searching for and using keys. Shelves near a door provide a place for people to sit packages while they negotiate the entrance of a door.

Lighting

Lighting is beneficial to all homeowners, but extra lighting might be needed for some people. Good lighting helps to prevent accidents and makes entry into a home much easier when natural light is minimal. Lights equipped with motion sensors are ideal, as they turn on automatically and require no effort on the part of the homeowner when approaching a dark house.

TRAFFIC CORRIDORS

Traffic corridors (Figure 24.3), such as halls and stairs should be wide and easily traversed. If a wheelchair is required for a homeowner to get around, all doors and corridors should have a minimum width of 3 feet; more is better. Be

---

- All access doors should be 3 feet wide.
- Push-button entry systems eliminate the need for fumbling with keys to door handles.
- Install lever-type door handles.
- Make sure good lighting is available at all access points.
- Shelving around an entry way can provide a place for people to set down packages while negotiating an entrance.
- Remote control of house lighting makes getting in and out of a home easier.
- All ramps, stoops, and steps should be slip-resistant and in good repair.
- Stair rails should be of a heavy-duty type that will support the full weight of a person who is struggling to gain good footing.

FIGURE 24.2 Access points.

- Doors must provide easy access.
- Bi-fold and sliding doors are less troublesome during access than a swinging door is.
- Use roll-up shades for window coverings. Multi-layered window dressings can be difficult for some people to deal with.
- Cabinet doors should be equipped with "D" shaped hardware so that a person can grip and use the handle easily.
- Entry doors should be equipped with lever-type handles. Round handles can pose problems for some people.
- Make sure that all doors and windows operate smoothly.

FIGURE 24.3 Building ideas for accommodating physically challenged people.

mindful of thresholds and install trim that makes transitions over thresholds easy and safe. If doors are installed in pathways, use sliding or folding doors whenever possible. Swinging doors can be difficult for some people to open and close. Also, lever handles on doors will make opening and closing the doors much easier than traditional round knobs. There are even add-on devices that install over common door knobs to make them of a lever type that is easy to operate.

STAIRS

Stairs that meet basic code requirements might be too small to provide safe passage for some people. Building stairs with lower risers and wider treads can make the use of them both easier and safer. Some stairs should be equipped with a power lift. An installation for a lift will require sturdy attachment and might require modification to existing framing in the wall that the lift will be mounted to. Handrails along stairs should be strong and attached with long screws that will not pull out under stress.

BATHROOMS

Bathrooms (Figure 24.4) in existing housing can present quite a challenge for a contractor who is required to make the room suitable for use from a wheel-

- Bathrooms for people with physical challenges often have to be larger than typical bathrooms.
- Use wall-mounted lavatories to facilitate the use of a wheelchair.
- Wall-mounted toilets can be installed at any height to accommodate the needs of a customer.
- Grab bars should be installed around the toilet, tub, and shower areas.
- Many plumbing fixtures are now available to allow easier access for people who do not have a full range of mobility.
- Ventilation is needed to eliminate steam and moisture that can cause a slipping hazard.
- All faucets should have bar-type handles for easy use.
- Floor-mounted toilets with taller profiles can make life easier for people who can't bend well.
- Make sure that the hot water is regulated not to exceed 110 degrees F.
- Pressure-balanced faucets should be used to prevent extreme variations in water temperatures.
- Your customer's condition and needs must be considered when installing mirrors, counters, and other options that can be installed at various heights.

FIGURE 24.4 Bathroom ideas.

chair. Most bathrooms are far too small to allow the use of a wheelchair. The first aspect of an existing bathroom that will probably need attention is the doorway. A wide, at least 3 feet, pocket door is an ideal solution to the tiny, swinging door that is likely to be present.

The floor surface in a bathroom should be impervious to water, but it should not become slippery when wet. A texture tile is a good choice when providing a finish floor covering in a bathroom. Some individuals might find commercial-grade carpeting to be a desirable covering when placed over a tile or vinyl flooring.

Design Considerations

There are many design considerations to take into account when building or remodeling a bathroom for individuals with physical challenges. Plumbing code requirements can be found in chapter 16.

Ideally, a bathroom laid out for a wheelchair should have an open turning area of at least 5 feet. Standard vanities are convenient for storage, but they can cause access problems for some people. A wall-mounted lavatory is often a better choice. Having a raised platform along the edge of a bathtub can make it easier for people to get into the tub. This is especially true of individuals who must transfer from a wheelchair, crutches, or a walker. Having a built-in, or after-market, seat in a shower or tub can also be very helpful for some people. A hand-held shower, on a flexible hose, is a good addition to most tubs and showers. Toilets should be of a tall, about 18 inches, type with an elongated bowl.

Grab bars are not only code requirements in handicap facilities, they are a darned good safety device in any bathroom. Comply with local code requirements on the design and placement of grab bars, but don't be shy to add more than the minimum number required. Adjustable guard rails for toilets are also a sensible installation in some cases.

Faucets should have lever-type handles, in either single-handle or two-handle design. Bathtubs and showers should be equipped with pressure-balanced, temperature-controlled fill valves to reduce the risk of scalding injuries and abrupt temperature changes. It is wise to keep hot water at a temperature that does not exceed 115 degrees F.

KITCHENS

Like bathrooms, kitchens can have you scratching your head when making conversions for people with physical restrictions (Figure 24.5). Establishing a counter height is one concern. If more than one person will be using the kitchen, the use of multilevel counters is a good idea. A counter that is user-friendly for a person in a wheelchair will be too low for a person who is not sitting in a chair. To combat this problem, install counters at varying heights that will accommodate all the cooks in the kitchen.

- Establish a counter height that will suit your customer's needs.
- Remember to leave leg room under countertops when your customer relies on a wheelchair for movement.
- Install pull-out shelves in cabinets.
- Pull-out and swing-out organizers can be used under counters and still allow adequate leg room.
- Wall-mounted ovens might be more beneficial to some customers than a floor unit would be.
- Cooktops should have staggered burners, so that residents will not have to reach across a hot burner to access a back burner.
- Lighting should be considered a key issue in a kitchen; provide plenty of it.
- Ventilation is always needed in a kitchen, but some cooks might require more clean air than others. Consider installing wall fans or other devices to keep a kitchen from becoming difficult to breath in.
- Fire extinguishers should be mounted at strategic locations throughout the kitchen.

FIGURE 24.5 Kitchen ideas.

Flooring

- Carpeting is a traditional flooring, but it might impede the use of a wheelchair or walker; consider using a vinyl floor covering in such instances. However, crutches might slip on vinyl, so carpeting might be used in areas where vinyl would normally be found.

Closets

- Bi-fold or sliding doors are the best options for closets.
- Walk-in closets should provide ample room for maneuvering with a wheelchair, walker, or crutches.
- Shelf height must be taken into consideration.
- All closets should be lighted.
- Closet poles must be set at appropriate heights.
- Utilize closet organizers to make the use of a closet easier.
- Paint the interior walls of closets with a paint that will reflect light well, therefore, providing better visibility.

FIGURE 24.6 Ideas for flooring and closets.

Equip cabinets with pull-out shelves and organizers. Rotating shelves in cabinets are also a good choice when creating easy access to kitchen supplies. For wheelchair use, you might have to install counters on cleats, rather than on base cabinets. Ranges and microwave ovens must be made accessible. This often means using a cooktop that is counter mounted and a separate oven that is wall mounted, rather than a combination oven and cooktop. A side-by-side refrigerator-freezer is, by far, the best choice, due to its low shelves in both the refrigerator and freezer compartments. Keep in mind, appliances are available with Braille control overlays for people who have trouble seeing.

CLOSETS

Closets (Figure 24.6) should be equipped with lights. Closet organizers are available in a wide range of configurations, and these devices can make a closet space more efficient and more user-friendly. Bi-fold, pocket, and sliding doors are the three preferred types of access for closets. Modification of shelf and closet-pole height might be required, and motorized carousels can prove beneficial. Pull-outs and roll-outs are both good ideas when making a closet more accessible.

DIFFERENCES

There are many differences in the needs of various people. While one person might need rocker-type electrical switches to overcome arthritic hands, another homeowner might require all vinyl flooring to facilitate the use of a wheelchair. A person who needs a shower built to accept a wheelchair might have no need for a microwave with a Braille overlay control panel. To make the most of your business and to create comfortable living conditions for your customers, you must assess each job on its own basis.

APPENDIX A
CONVERSION TABLES

To change	To	Multiply by
Inches	Feet	0.0833
Inches	Millimeters	25.4
Feet	Inches	12
Feet	Yards	0.3333
Yards	Feet	3
Square inches	Square feet	0.00694
Square feet	Square inches	144
Square feet	Square yards	0.11111
Square yards	Square feet	9
Cubic inches	Cubic feet	0.00058
Cubic feet	Cubic inches	1728
Cubic feet	Cubic yards	0.03703
Gallons	Cubic inches	231
Gallons	Cubic feet	0.1337
Gallons	Pounds of water	8.33
Pounds of water	Gallons	0.12004
Ounces	Pounds	0.0625
Pounds	Ounces	16
Inches of water	Pounds per square inch	0.0361
Inches of water	Inches of mercury	0.0735
Inches of water	Ounces per square inch	0.578
Inches of water	Pounds per square foot	5.2
Inches of mercury	Inches of water	13.6
Inches of mercury	Feet of water	1.1333
Inches of mercury	Pounds per square inch	0.4914
Ounces per square inch	Inches of mercury	0.127
Ounces per square inch	Inches of water	1.733
Pounds per square inch	Inches of water	27.72
Pounds per square inch	Feet of water	2.310
Pounds per square inch	Inches of mercury	2.04
Pounds per square inch	Atmospheres	0.0681
Feet of water	Pounds per square inch	0.434
Feet of water	Pounds per square foot	62.5
Feet of water	Inches of mercury	0.8824

FIGURE A.1 Measurement conversion factors.

FIGURE A.1 Measurement conversion factors *continued.*

To change	To	Multiply by
Atmospheres	Pounds per square inch	14.696
Atmospheres	Inches of mercury	29.92
Atmospheres	Feet of water	34
Long tons	Pounds	2240
Short tons	Pounds	2000
Short tons	Long tons	0.89295

To find	Multiply	By
Microns	Mils	25.4
Centimeters	Inches	2.54
Meters	Feet	0.3048
Meters	Yards	0.19144
Kilometers	Miles	1.609344
Grams	Ounces	28.349523
Kilograms	Pounds	0.4539237
Liters	Gallons (U.S.)	3.7854118
Liters	Gallons (Imperial)	4.546090
Milliliters (cc)	Fluid ounces	29.573530
Milliliters (cc)	Cubic inches	16.387064
Square centimeters	Square inches	6.4516
Square meters	Square feet	0.09290304
Square meters	Square yards	0.83612736
Cubic meters	Cubic feet	2.8316847×10^{-2}
Cubic meters	Cubic yards	0.76455486
Joules	BTU	1054.3504
Joules	Foot-pounds	1.35582
Kilowatts	BTU per minute	0.01757251
Kilowatts	Foot-pounds per minute	2.2597×10^{-5}
Kilowatts	Horsepower	0.7457
Radians	Degrees	0.017453293
Watts	BTU per minute	17.5725

FIGURE A.2 Conversion factors in converting from customary (U.S.) units to metric units.

	Imperial	Metric
Length	1 inch	25.4 mm
	1 foot	0.3048 m
	1 yard	0.9144 m
	1 mile	1.609 km
Mass	1 pound	0.454 kg
	1 U.S. short ton	0.9072 tonne
Area	1 ft^2	0.092 m^2
	1 yd^2	0.836 m^2
	1 acre	0.404 hectare (ha)
Capacity/Volume	1 ft^3	0.028 m^3
	1 yd^3	0.764 m^3
	1 liquid quart	0.946 litre (1)
	1 gallon	3.785 litre (1)
Heat	1 BTU	1055 joule (J)
	1 BTU/hr	0.293 watt (W)

FIGURE A.3 Measurement conversions: Imperial to metric.

Volume	Weight
1 cu. ft. sand	Approx. 100 lbs.
1 cu. yd.	2700 lbs.
1 ton	¾ yd. or 20 cu. ft.
Avg. shovelful	15 lbs.
12 qt. pail	40 lbs.

FIGURE A.4 Sand volume to weight conversions.

U.S.	Metric
0.001 inch	0.025 mm
1 inch	25.400 mm
1 foot	30.48 cm
1 foot	0.3048 m
1 yard	0.9144 m
1 mile	1.609 km

FIGURE A.5 Conversion tables.

FIGURE A.5 Conversion tables *continued.*

U.S.	Metric
1 inch^2	6.4516 cm^2
1 feet^2	0.0929 m^2
1 yard^2	0.8361 m^2
1 acre	0.4047 ha
1 mile^2	2.590 km^2
1 inch^3	16.387 cm^3
1 feet^3	0.0283 m^3
1 yard^3	0.7647 m^3
1 U.S. ounce	29.57 ml
1 U.S. pint	0.4732 l
1 U.S. gallon	3.785 l
1 ounce	28.35 g
1 pound	0.4536 kg

Unit	Equals
1 meter	39.3 inches 3.28083 feet 1.0936 yards
1 centimeter	.3937 inch
1 millimeter	.03937 inch, or nearly ⅖ inch
1 kilometer	0.62137 mile
1 foot	.3048 meter
1 inch	2.54 centimeters 25.40 millimeters
1 square meter	10.764 square feet 1.196 square yards
1 square centimeter	.155 square inch
1 square millimeter	.00155 square inch
1 square yard	.836 square meter
1 square foot	.0929 square meter
1 square inch	6.452 square centimeter 645.2 square millimeter

FIGURE A.6 Metric-customary equivalents.

Unit	Equals
1 cubic meter	35.314 cubic feet 1.308 cubic yards 264.2 U.S. gallons (231 cubic inches)
1 cubic decimeter	61.0230 cubic inches .0353 cubic feet
1 cubic centimeter	.061 cubuic inch
1 liter	1 cubic decimeter 61.0230 cubic inches 0.0353 cubic foot 1.0567 quarts (U.S.) 0.2642 gallon (U.S.) 2.2020 lb. of water at 62°F.
1 cubic yard	.7645 cubic meter
1 cubic foot	.02832 cubic meter 28.317 cubic decimeters 28.317 liters
1 cubic inch	16.383 cubic centimeters
1 gallon (British)	4.543 liters
1 gallon (U.S.)	3.785 liters
1 gram	15.432 grains
1 kilogram	2.2046 pounds
1 metric ton	.9842 ton of 2240 pounds 19.68 cwts. 2204.6 pounds
1 grain	.0648 gram
1 ounce avoirdupois	28.35 grams
1 pound	.4536 kilograms
1 ton of 2240 lb.	1.1016 metric tons 1016 kilograms

FIGURE A.7 Measures of volume and capacity.

Quantity	Unit	Symbol
Time	Second	s
Plane angle	Radius	rad
Force	Newton	N
Energy, work, quantity of heat	Joule Kilojoule Megajoule	J kJ MJ
Power, heat flow rate	Watt Kilowatt	W kW
Pressure	Pascal Kilopascal Megapascal	Pa kPa MPa
Velocity, speed	Meter per second Kilometer per hour	m/s km/h

FIGURE A.8 Metric symbols.

Inches^2	Millimeters^2
0.01227	8.0
0.04909	31.7
0.11045	71.3
0.19635	126.7
0.44179	285.0
0.7854	506.7
1.2272	791.7
1.7671	1140.1
3.1416	2026.8
4.9087	3166.9
7.0686	4560.4
12.566	8107.1
19.635	12667.7
28.274	18241.3
38.485	24828.9
50.265	32478.9
63.617	41043.1
78.540	50670.9

FIGURE A.9 Area in inches and millimeters.

Inches	Millimeters
0.3927	10
0.7854	20
1.1781	30
1.5708	40
2.3562	60
3.1416	80
3.9270	100
4.7124	120
6.2832	160
7.8540	200
9.4248	240
12.566	320
15.708	400
18.850	480
21.991	560
25.133	640
28.274	720
31.416	800

FIGURE A.10 Circumference in inches and millimeters.

Inches^2	Millimeters^2
0.01227	8.0
0.04909	31.7
0.11045	71.3
0.19635	126.7
0.44179	285.0
0.7854	506.7
1.2272	791.7
1.7671	1140.1
3.1416	2026.8
4.9087	3166.9
7.0686	4560.4

FIGURE A.11 Area in inches and millimeters.

FIGURE A.11 Area in inches and millimeters *continued.*

Inches^2	Millimeters^2
12.566	8107.1
19.635	12667.7
28.274	18241.3
38.485	24828.9
50.265	32478.9
63.617	41043.1
78.540	50670.9

Feet	Meters (m)	Millimeters (mm)
1	0.305	304.8
2	0.610	609.6
3 (1 yd.)	0.914	914.4
4	1.219	1 219.2
5	1.524	1 524.0
6 (2 yd.)	1.829	1 828.8
7	2.134	2 133.6
8	2.438	2 438.2
9 (3yd.)	2.743	2 743.2
10	3.048	3 048.0
20	6.096	6 096.0
30 (10 yd.)	9.144	9 144.0
40	12.19	12 192.0
50	15.24	15 240.0
60 (20 yd.)	18.29	18 288.0
70	21.34	21 336.0
80	24.38	24 384.0
90 (30 yd.)	27.43	27 432.0
100	30.48	30 480.0

FIGURE A.12 Length conversions.

Units	Equals
1 decimeter	4 inches
1 meter	1.1 yards
1 kilometer	⅝ mile
1 hektar	2½ acres
1 stere or cu. meter	¼ cord
1 liter	1.06 qt. liquid; 0.9 qt. dry
1 hektoliter	2⅝ bushel
1 kilogram	2⅕ lbs.
1 metric ton	2200 lbs.

FIGURE A.13 Approximate metric equivalents.

Inches	Millimeters
1	25.4
2	50.8
3	76.2
4	101.6
5	127.0
6	152.4
7	177.8
8	203.2
9	228.6
10	254.0
11	279.4
12	304.8
13	330.2
14	355.6
15	381.0
16	406.4
17	431.8
18	457.2
19	482.6
20	508.0

FIGURE A.14 Inches to millimeters.

Metric linear measure		
Measure	**Equals**	**Equals**
	1 millimeter	.001 meter
10 millimeter	1 centimeter	.01 meter
10 centimeter	1 decimeter	.1 meter
10 decimeter	1 meter	1 meter
10 meters	1 dekameter	10 meters
10 dekameters	1 hectometer	100 meters
10 hectometers	1 kilometer	1000 meters
10 kilometers	1 myriameter	10,000 meters

Metric land measure	
Unit	**Equals**
1 centiare (ca.)	1 sq. meter
100 centiares (ca.)	1 are
100 ares (A.)	1 hectare
100 hectares (ha.)	1 sq. kilometer

FIGURE A.15 Metric linear measurements.

Inches	**Meters (m)**	**Millimeters (mm)**
⅛	0.003	3.17
¼	0.006	6.35
⅜	0.010	9.52
½	0.013	12.6
⅝	0.016	15.87
¾	0.019	19.05
⅞	0.022	22.22
1	0.025	25.39
2	0.051	50.79
3	0.076	76.20
4	0.102	101.6
5	0.127	126.9
6	0.152	152.4
7	0.178	177.8

FIGURE A.16 Length conversions.

FIGURE A.16   Length conversions *continued.*

Inches	Meters (m)	Millimeters (mm)
8	0.203	203.1
9	0.229	228.6
10	0.254	253.9
11	0.279	279.3
12	0.305	304.8

Quantity	Unit	Symbol
Length	Millimeter	mm
	Centimeter	cm
	Meter	m
	Kilometer	km
Area	Square Millimeter	mm^7
	Square Centimeter	cm^2
	Square Decimeter	dm^2
	Square Meter	m^2
	Square Kilometer	km^2
Volume	Cubic Centimeter	cm^3
	Cubic Decimeter	dm^3
	Cubic Meter	m^3
Mass	Milligram	mg
	Gram	g
	Kilogram	kg
	Tonne	t
Temperature	Degree Celsius	°C
	Kelvin	K
Time	Second	s
Plane angle	Radius	rad
Force	Newton	N
Energy, work, quantity of heat	Joule	J
	Kilojoule	kJ
	Megajoule	MJ
Power, heat flow rate	Watt	W
	Kilowatt	kW
Pressure	Pascal	Pa
	Kilopascal	kPa
	Megapascal	MPa

FIGURE A.17   Metric symbols.

FIGURE A.17 Metric symbols *continued.*

Quantity	Unit	Symbol
Velocity, speed	Meter per second	m/s
	Kilometer per hour	km/h
Revolutional frequency	Revolution per minute	r/min

1 cu. ft. at 50°F. weighs 62.41 lb.
1 gal. at 50°F weighs 8.34 lb.
1 cu. ft. of ice weighs 57.2 lb.
Water is at its greatest density at 39.2°F.
1 cu. ft. at 39.2°F. weighs 62.43 lb.

FIGURE A.18 Water weight.

Quantity	Equals
10 milligrams (mg.)	1 centigram
10 centigrams (cg.)	1 decigram
10 decigrams (dg.)	1 gram
10 grams (g.)	1 dekagram
10 dekagrams (Dg.)	1 hectogram
10 hectograms (hg.)	1 kilogram
10 kilogram (kg.)	1 myriagram
10 myriagrams (Mg.)	1 quintal

FIGURE A.19 Metric weight measure.

Square feet	Square meters
1	0.925
2	.1850
3	.2775
4	.3700
5	.4650
6	.5550
7	.6475
8	.7400
9	.8325
10	.9250
25	2.315
50	4.65
100	9.25

FIGURE A.20 Square feet to square meters.

Quantity	Equals
Metric cubic measure	
1000 cubic millimeters (cu. mm.)	1 cubic centimeter
1000 cubic centimeters (cu. cm.)	1 cubic decimeter
1000 cubic decimeters (cu. dm.)	1 cubic meter
Metric capacity measure	
10 milliliters (mi.)	1 centiliter
10 centiliters (cl.)	1 deciliter
10 deciliters (dl.)	1 liter
10 liters (l.)	1 dekaliter
10 dekaliters (Dl.)	1 hectoliter
10 hectoliters (hl.)	1 kiloliter
10 kiloliters (kl.)	1 myrialiter (ml.)

FIGURE A.21 Metric cubic measure.

To change	To	Multiply by
Inches	Millimeters	25.4
Feet	Meters	.3048
Miles	Kilometers	1.6093
Square inches	Square centimeters	6.4515
Square feet	Square meters	.09290
Acres	Hectares	.4047
Acres	Square kilometers	.00405
Cubic inches	Cubic centimeters	16.3872
Cubic feet	Cubic meters	.02832
Cubic yards	Cubic meters	.76452
Cubic inches	Liters	.01639
U.S. gallons	Liters	3.7854
Ounces (avoirdupois)	Grams	28.35
Pounds	Kilograms	.4536
Lbs. per sq. in. (P.S.I.)	Kg.'s per sq. cm.	.0703
Lbs. per cu. ft.	Kg.'s per cu. meter	16.0189
Tons (2000 lbs.)	Metric tons (1000 kg.)	.9072
Horsepower	Kilowatts	.746

FIGURE A.22 English to metric conversions.

Quantity	Equals
100 sq. millimeters	1 sq. centimeter
100 sq. centimeters	1 sq. decimeter
100 sq. decimeters	1 sq. meter

FIGURE A.23 Metric square measure.

Quantity	Equals
1 meter	39.3 inches 3.28083 feet 1.0936 yards
1 centimeter	.3937 inch
1 millimeter	.03937 inch, or nearly 1/25 inch
1 kilometer	0.62137 mile
.3048 meter	1 foot
2.54 centimeters	1 inch 25.40 millimeters

FIGURE A.24 Metric conversions.

Quantity	Equals	Equals
12 inches	1 foot	
3 feet	1 yard	36 inches
5½ yards	1 rod	16½ feet
40 rods	1 furlong	660 feet
8 furlongs	1 mile	5280 feet

FIGURE A.25 Linear measure.

Quantity	Equals
Linear measure	
12 inches	1 foot
3 feet	1 yard
5½ yards	1 rod
320 rods	1 mile
1 mile	1760 yards
1 mile	5280 feet
Square measure	
144 sq. inches	1 sq. foot
9 sq. feet	1 sq. yard
1 sq. yard	1296 sq. inches
4840 sq. yards	1 acre
640 acres	1 sq. mile

FIGURE A.26 Weights and measures.

FIGURE A.26 Weights and measures *continued.*

Quantity	Equals
Cubic measure	
1728 cubic inches	1 cubic foot
27 cubic feet	1 cubic yard
Avoirdupois weight	
16 ounces	1 pound
100 pounds	1 hundredweight
20 hundredweight	1 ton
1 ton	2000 pounds
1 long ton	2240

Quantity	Equals
4 gills	1 pint
2 pints	1 quart
4 quarts	1 gallon
31½ gallons	1 barrel
1 gallon	231 cubic inches
7.48 gallons	1 cubic foot
1 gallon water	8.33 pounds
1 gallon gasoline	5.84 pounds

FIGURE A.27 Liquid measure.

Unit	Equals
1 cu. ft.	62.4 lbs.
1 cu. ft.	7.48 gal.
1 gal.	8.33 lbs.
1 gal.	0.1337 cu. ft.

FIGURE A.28 Water volume to weight conversion.

Unit	Equals
1 gallon	0.133681 cubic foot
1 gallon	231 cubic inches

FIGURE A.29 Volume measure equivalents.

Quantity	Equals
12 inches (in. or ")	1 foot (ft. or ')
3 feet	1 yard (yd.)
5½ yards or 16½ feet	1 rod (rd.)
40 rods	1 furlong (fur.)
8 furlongs or 320 rods	1 statute mile (mi.)

FIGURE A.30 Long measure.

Unit	Equals
1 sq. centimeter	0.1550 sq. in.
1 sq. decimeter	0.1076 sq. ft.
1 sq. meter	1.196 sq. yd.
1 are	3.954 sq. rd.
1 hektar	2.47 acres
1 sq. kilometer	0.386 sq. mile
1 sq. in.	6.452 sq. centimeters
1 sq. ft.	9.2903 sq. decimeters
1 sq. yd.	0.8361 sq. meter
1 sq. rd.	0.2529 are
1 acre	0.4047 hektar
1 sq. mile	2.59 sq. kilometers

FIGURE A.31 Square measures.

Quantity	Equals
1 square meter	10.764 square feet
	1.196 square yards
1 square centimeter	.155 square inch
1 square millimeter	.00155 square inch
.836 square meter	1 square yard
.0929 square meter	1 square foot
6.452 square centimeter	1 square inch
645.2 square millimeter	1 square inch

FIGURE A.32 Surface measures.

Quantity	Equals	Equals
7.92 inches	1 link	
100 links	1 chain	66 feet
10 chains	1 furling	660 feet
80 chains	1 mile	5280 feet

FIGURE A.33 Surveyor's measure.

Quantity	Equals
144 sq. inches	1 sq. foot
9 sq. feet	1 sq. yard
1 sq. yard	1296 sq. inches
4840 sq. yards	1 acre
640 acres	1 sq. mile

FIGURE A.34 Square measure.

Inches	Decimals of an inch
1/16	.0156
1/32	.0312
3/64	.0468
1/16	.0625
5/64	.0781
3/32	.0937
7/64	.1093
1/8	.1250
9/64	.1406
5/32	.1562
11/64	.1718
3/16	.1875
13/64	.2031
7/32	.2187
15/64	.2343
1/4	.2500
17/64	.2656
9/32	.2812
19/64	.2968
5/16	.3125

FIGURE A.35 Decimal equivalents of fractions.

Inches	Millimeters
1/8	3.2
1/4	6.4
3/8	9.5
1/2	12.7
3/4	19.1
1	25.4
1 1/4	31.8
1 1/2	38.1
2	50.8
2 1/2	63.5

FIGURE A.36 Diameter in inches and millimeters.

FIGURE A.36 Diameter in inches and millimeters *continued.*

Inches	Millimeters
3	76.2
4	101.6
5	127
6	152.4
7	177.8
8	203.2
9	228.6
10	254

Unit	Equals
1 gram	15.432 grains
1 kilogram	2.2046 pounds
1 metric ton	.9842 ton of 2240 pounds 19.68 cwts. 2204.6 pounds
1 grain	.0648 gram
1 ounce avoirdupois	28.35 grams
1 pound	.4536 kilograms
1 ton of 2240 lb.	1.1016 metric tons 1016 kilograms

FIGURE A.37 Weight conversions.

Quantity	Equals	Meters	English equivalent
1 mm.	1 millimeter	1/1000	.03937 in.
10 mm.	1 centimeter	1/100	.3937 in.
10 cm.	1 decimeter	1/10	3.937 in.
10 dm.	1 meter	1	39.37 in.
10 m.	1 dekameter	10	32.8 ft.
10 Dm.	1hectometer	100	328.09 ft.
10 Hm.	1 kilometer	1000	.62137 mile

FIGURE A.38 Lengths.

Quantity	Equals	Equals
144 sq. inches	1 sq. foot	
9 sq. feeet	1 sq. yard	
30¼ sq. yards	1 sq. rod	272.25 sq. feet
160 sq. rods	1 acre	4840 sq. yards or 43,560 sq. feet
640 acres	1 sq. mile	3,097,600 sq. yards
36 sq. miles	1 township	

FIGURE A.39   Square measure.

Quantity	Equals	Cubic inches
2 pints	1 quart	67.2
8 quarts	1 peck	537.61
4 pecks	1 bushel	2150.42

FIGURE A.40   Dry measures.

Quantity	Equals
144 sq. in.	1 sq. ft.
9 sq. ft.	1 sq. yd.
30½ sq. yd.	1 sq. rd.
160 sq. rd.	1 acre
640 acres	1 sq. mile
43,560 sq. ft.	1 acre

FIGURE A.41   Surface measure.

Passageway	Recommended	Minimum
Stairs	40"	36"
Landings	40"	36"
Main hall	48"	36"
Minor hall	36"	30"
Interior door	32"	28"
Exterior door	36"	36"

FIGURE A.42 Widths of passageways.

Barometer (ins. of mercury)	Pressure (lbs. per sq. in.)
28.00	13.75
28.25	13.88
28.50	14.00
28.75	14.12
29.00	14.24
29.25	14.37
29.50	14.49
29.75	14.61
29.921	14.696
30.00	14.74
30.25	14.86
30.50	14.98
30.75	15.10
31.00	15.23

Rule: Barometer in inches of mercury × .49116 = lbs. per sq. in.

FIGURE A.43 Atmospheric pressure per square inch.

APPENDIX B
HELPFUL CALCULATION METHODS

(Surface area ÷ R value) × (temperature inside – temperature outside)

Surface area of a material (in square feet) divided by its "R" value and multiplied by the difference in Fahrenheit degrees between inside and outside temperature equals heat loss in BTU's per hour.

FIGURE B.1   Calculating heat loss per hour with R-value.

- 3 feet of 1-in. pipe equal 1 square foot of radiation.
- 2⅓ linear feet of 1¼ in. pipe equal 1 square foot of radiation.
- Hot water radiation gives off 150 BTU per square foot of radiation per hour.
- Steam radiation gives off 240 BTU per square foot of radiation per hour.
- On greenhouse heating, figure ⅔ square foot of radiation per square foot of glass.
- One square foot of direct radiation condenses .25 pound of water per hour.

FIGURE B.2   Radiant heat facts.

$$L = \frac{144}{D \times 3.1414} \times R \div 12$$

D = O.D. of pipe
L = length of pipe needed in ft.
R = sq. ft. of radiation needed

FIGURE B.3   Formulas for pipe radiation.

The approximate weight of a piece of pipe can be determined by the following formulas:

Cast Iron Pipe: weight = $(A^2 - B^2) \times C \times .2042$

Steel Pipe: weight = $(A^2 - B^2) \times C \times .2199$

Copper Pipe: weight = $(A^2 - B^2) \times C \times .2537$

A = outside diameter of the pipe in inches

B = inside diameter of the pipe in inches

C = length of the pipe in inches

FIGURE B.4   Finding the weight of piping.

The area of a pipe wall can be determined by the following formula:
Area of pipe wall = .7854 × [(O.D. × O.D.) − (I.D. × I.D.)]

FIGURE B.5 Finding the area of a pipe.

The formula for calculating expansion or contraction in plastic piping is:

$$L = Y \times \frac{T - F}{10} \times \frac{L}{100}$$

L = Expansion in inches
Y = Constant factor expressing inches of expansion per 100°F temperature change per 100 ft. of pipe
T = Maximum temperature (0°F)
F = Minimum temperature (0°F)
L = Length of pipe run in feet

FIGURE B.6 Expansion in plastic piping.

The capacity of pipes is as the square of their diameters. Thus, doubling the diameter of a pipe increases its capacity four times.

FIGURE B.7 A piping fact.

Inch scale	Metric scale
¹⁄₁₆"	1:200

FIGURE B.8 Scale used for site plans.

Inch scale	Metric scale
¼"	1:50
⅛"	1:100

FIGURE B.9 Scales used for building plans.

To change	To	Multiply by
Inches	Feet	0.0833
Inches	Millimeters	25.4
Feet	Inches	12
Feet	Yards	0.3333
Yards	Feet	3
Square inches	Square feet	0.00694
Square feet	Square inches	144
Square feet	Square yards	0.11111
Square yards	Square feet	9
Cubic inches	Cubic feet	0.00058
Cubic feet	Cubic inches	1728
Cubic feet	Cubic yards	0.03703
Cubic yards	Cubic feet	27
Cubic inches	Gallons	0.00433
Cubic feet	Gallons	7.48
Gallons	Cubic inches	231
Gallons	Cubic feet	0.1337
Gallons	Pounds of water	8.33
Pounds of water	Gallons	0.12004
Ounces	Pounds	0.0625
Pounds	Ounces	16
Inches of water	Pounds per square inch	0.0361
Inches of water	Inches of mercury	0.0735
Inches of water	Ounces per square inch	0.578
Inches of water	Pounds per square foot	5.2
Inches of mercury	Inches of water	13.6
Inches of mercury	Feet of water	1.1333
Inches of mercury	Feet of water	0.4914
Ounces per square inch	Pounds per square inch	0.127
Ounces per square inch	Inches of mercury	1.733
Pounds per square inch	Inches of water	27.72
Pounds per square inch	Feet of water	2.310
Pounds per square inch	Inches of mercury	2.04
Pounds per square inch	Atmospheres	0.0681

FIGURE B.10 Useful multipliers.

FIGURE B.10 Useful multipliers *continued.*

To change	To	Multiply by
Feet of water	Pounds per square inch	0.434
Feet of water	Pounds per square foot	62.5
Feet of water	Inches of mercury	0.8824
Atmospheres	Pounds per square inch	14.696
Atmospheres	Inches of mercury	29.92
Atmospheres	Feet of water	34
Long tons	Pounds	2240
Short tons	Pounds	2000
Short tons	Long tons	0.89295

To figure the final temperature when two different temperatures of water are mixed together, use the following formula:

$$\frac{(A \times C) + (B \times D)}{A + B}$$

A = Weight of lower temperature water
B = Weight of higher temperature water
C = Lower temperature
D = Higher temperature

FIGURE B.11 Temperature calculation.

Temperature can be expressed according to the Fahrenheit scale or the Celsius scale. To convert C to F or F to C, use the following formulas:

$$°F = 1.8 \times °C + 32$$
$$°C = 0.55555555 \times °F - 32$$

FIGURE B.12 Temperature conversion.

Deg. C. = Deg. F. − 32 ÷ 1.8

FIGURE B.13 Temperature conversion.

$$\text{Deg. F.} = \text{Deg. C.} \times 1.8 + 32$$

FIGURE B.14 Temperature conversion.

Outside design temperature = Average of lowest recorded temperature in
each month from October to March.
Inside design temperature = 70° Fahrenheit or as specified by owner.

A degree day is one day × the number of Fahrenheit degrees the mean
temperature is below 65°F. The number of degree days in a year is a good
guideline for designing heating and insulation systems.

FIGURE B.15 Design temperature.

C	Base temperature	F
−73	−100	−148
−68	−90	−130
−62	−80	−112
−57	−70	−94
−51	−60	−76
−46	−50	−58
−40	−40	−40
−34.4	−30	−22
−28.9	−20	−4
−23.3	−10	14
−17.8	0	32
−17.2	1	33.8
−16.7	2	35.6
−16.1	3	37.4
−15.6	4	39.2
−15.0	5	41.0
−14.4	6	42.8
−13.9	7	44.6
−13.3	8	46.4
−12.8	9	48.2
−12.2	10	50.0

FIGURE B.16 Temperature conversion: −100 to 30.

FIGURE B.16 Temperature conversion: −100 to 30 *continued.*

C	Base temperature	F
−11.7	11	51.8
−11.1	12	53.6
−10.6	13	55.4
−10.0	14	57.2

C	Base temperature	F
−0.6	31	87.8
0	32	89.6
0.6	33	91.4
1.1	34	93.2
1.7	35	95.0
2.2	36	96.8
2.8	37	98.6
3.3	38	100.4
3.9	39	102.2
4.4	40	104.0
5.0	41	105.8
5.6	42	107.6
6.1	43	109.4
6.7	44	111.2
7.2	45	113.0
7.8	46	114.8
8.3	47	116.6
8.9	48	118.4
9.4	49	120.0
10.0	50	122.0
10.6	51	123.8
11.1	52	125.6
11.7	53	127.4
12.2	54	129.2
12.8	55	131.0

FIGURE B.17 Temperature conversion: 31 to 71.

C	Base temperature	F
22.2	72	161.6
22.8	73	163.4
23.3	74	165.2
23.9	75	167.0
24.4	76	168.8
25.0	77	170.6
25.6	78	172.4
26.1	79	174.2
26.7	80	176.0
27.8	81	177.8
28.3	82	179.6
28.9	83	181.4
29.4	84	183.2
30.0	85	185.0
30.6	86	186.8
31.1	87	188.6
31.7	88	190.4
32.2	89	192.2
32.8	90	194.0
33.3	91	195.8
33.9	92	197.6
34.4	93	199.4
35.0	94	201.2
35.6	95	203.0
36.1	96	204.8

FIGURE B.18 Temperature conversion: 72 to 212.

C	Base temperature	F
104	220	248
110	230	446
116	240	464
121	250	482
127	260	500
132	270	518
138	280	536
143	290	554
149	300	572
154	310	590
160	320	608
166	330	626
171	340	644
177	350	662
182	360	680
188	370	698
193	380	716
199	390	734
204	400	752
210	410	770
216	420	788
221	430	806
227	440	824
232	450	842
238	460	860

FIGURE B.19 Temperature conversion: 213 to 620.

C	Base temperature	F
332	630	1166
338	640	1184
343	650	1202
349	660	1220
354	670	1238
360	680	1256
366	690	1274
371	700	1292
377	710	1310
382	720	1328
388	730	1346
393	740	1364
399	750	1382
404	760	1400
410	770	1418
416	780	1436
421	790	1454
427	800	1472
432	810	1490
438	820	1508
443	830	1526
449	840	1544
454	850	1562
460	860	1580
466	870	1598

FIGURE B.20 Temperature conversion: 621 to 1000.

Quantity	Equals
60 seconds	1 minute
60 minutes	1 degree
360 degrees	1 circle

FIGURE B.21 Circular measure.

Function	Formula
Sine	$\sin = \dfrac{\text{side opposite}}{\text{hypotenuse}}$
Cosine	$\cos = \dfrac{\text{side adjacent}}{\text{hypotenuse}}$
Tangent	$\tan = \dfrac{\text{side opposite}}{\text{side adjacent}}$
Cosecant	$\csc = \dfrac{\text{hypotenuse}}{\text{side opposite}}$
Secant	$\sec = \dfrac{\text{hypotenuse}}{\text{side adjacent}}$
Cotangent	$\cot = \dfrac{\text{side adjacent}}{\text{side opposite}}$

FIGURE B.22 Trigonometry.

Multiply Length × Width × Thickness
Example: 50 ft. × 10 ft. × 8 in.
50' × 10' × ⁸⁄₁₂' = 333.33 cu. feet
To convert to cubic yards, divide by 27 cu. ft. per cu. yd.
Example: 333.33 ÷ 27 = 12.35 cu. yd.

FIGURE B.23 Estimating volume.

Area of surface = Diameter × 3.1416 × length + area of the two bases
Area of base = Diameter × diameter × .7854
Area of base = Volume ÷ length
Length = Volume ÷ area of base
Volume = Length × area of base
Capacity in gallons = Volume in inches ÷ 231
Capacity of gallons = Diameter × diameter × length × .0034
Capacity in gallons = Volume in feet × 7.48

FIGURE B.24 Cylinder formulas.

Area = Short diameter × long diameter × .7854

FIGURE B.25 Ellipse calculation.

Area of surface = One half of circumference of base × slant height + area
of base.
Volume = Diameter × diameter × .7854 × one-third of the altitude.

FIGURE B.26   Cone calculation.

Volume = Width × height × length

FIGURE B.27   Volume of a rectangular prism.

Area = Length × width

FIGURE B.28   Finding the area of a square.

Area = ½ perimeter of base × slant height + area of base
Volume = Area of base × ⅓ of the altitude

FIGURE B.29   Finding area and volume of a pyramid.

These comprise the numerous figures having more than four sides, names
according to the number of sides, thus:

Figure	**Sides**
Pentagon	5
Hexagon	6
Heptagon	7
Octagon	8
Nonagon	9
Decagon	10

To find the area of a polygon: Multiply the sum of the sides (perimeter of
the polygon) by the perpendicular dropped from its center to one of its
sides, and half the product will be the area. This rule applies to all regular
polygons.

FIGURE B.30   Polygons.

Area = Width of side × 2.598 × width of side

FIGURE B.31   Hexagons.

Area = Base × distance between the two parallel sides

FIGURE B.32 Parallelograms.

Area = Length × width

FIGURE B.33 Rectangles.

Area of surface = Diameter × diameter × 3.1416
Side of inscribed cube = Radius × 1.547
Volume = Diameter × diameter × diameter × .5236

FIGURE B.34 Spheres.

Area = One-half of height times base

FIGURE B.35 Triangles.

Area = One-half of the sum of the parallel sides × the height

FIGURE B.36 Trapezoids.

Volume = Width × height × length

FIGURE B.37 Cubes.

Circumference = Diameter × 3.1416
Circumference = Radius × 6.2832
Diameter = Radius × 2
Diameter = Square root of (area ÷ .7854)
Diameter = Square root of area × 1.1283
Diameter = Circumference × .31831
Radius = Diameter ÷ 2
Radius = Circumference × .15915
Radius = Square root of area × .56419
Area = Diameter × Diameter × .7854
Area = Half of the circumference × half of the diameter
Area = Square of the circumference × .0796
Arc length = Degrees × radius × .01745
Degrees of arc = Length ÷ (radius × .01745)
Radius of arc = Length ÷ (degrees × .01745)
Side of equal square = Diameter × .8862
Side of inscribed square = Diameter × .7071
Area of sector = Area of circle × degrees of arc ÷ 360

FIGURE B.38   Formulas for a circle.

1. Circumference of a circle = π × diameter or 3.1416 × diameter
2. Diameter of a circle = Circumference × .31831
3. Area of a square = Length × width
4. Area of a rectangle = Length × width
5. Area of a parallelogram = Base × perpendicular height
6. Area of a triangle = ½ base × perpendicular height
7. Area of a circle = π × radius squared or diameter squared × .7854
8. Area of an ellipse = Length × width × .7854
9. Volume of a cube or rectangular prism = Length × width × height
10. Volue of a triangular prism = Area of triangle × length
11. Volume of a sphere = Diameter cubed × .5236 or (dia. × dia. × dia. × .5236)
12. Volume of a cone = π × radius square × ⅓ height
13. Volume of a cylinder = π × radius squared × height
14. Length of one side of a square × 1.128 = Diameter of an equal circle
15. Doubling the diameter of a pipe or cylinder increases its capacity 4 times
16. The pressure (in lbs. per sq. inch) of a column of water = Height of the column (in feet) × .434
17. The capacity of a pipe or tank (in U.S. gallons) = Diameter squared (in inches) × the length (in inches) × .0034
18. A gallon of water = 8⅓ lb. = 231 cu. inches
19. A cubic foot of water = 62½ lb. = 7½ gallons

FIGURE B.39   Useful formulas.

Number	Square
1	1
2	4
3	9
4	16
5	25
6	36
7	49
8	64
9	81
10	100
11	121
12	144
13	169
14	196
15	225
16	256
17	289
18	324
19	361
20	400
21	441
22	484
23	529
24	576
25	625
26	676
27	729
28	784
29	841
30	900
31	961
32	1024
33	1089
34	1156
35	1225

FIGURE B.40 Squares of numbers.

FIGURE B.40 Squares of numbers *continued.*

Number	Square
36	1296
37	1369
38	1444
39	1521
40	1600
41	1681
42	1764
43	1849
44	1936
45	2025
46	2116
47	2209
48	2304
49	2401
50	2500
51	2601
52	2704
53	2809
54	2916
55	3025
56	3136
57	3249
58	3364
59	3481
60	3600
61	3721
62	3844
63	3969
64	4096
65	4225
66	4356
67	4489
68	4624
69	4761

FIGURE B.40 Squares of numbers *continued.*

Number	Square
70	4900
71	5041
72	5184
73	5329
74	5476
75	5625
76	5776
77	5929
78	6084
79	6241
80	6400
81	6561
82	6724
83	6889
84	7056
85	7225
86	7396
87	7569
88	7744
89	7921
90	8100
91	8281
92	8464
93	8649
94	8836
95	9025
96	9216
97	9409
98	9604
99	9801
100	10000

Number	Square root
1	1.00000
2	1.41421
3	1.73205
4	2.00000
5	2.23606
6	2.44948
7	2.64575
8	2.82842
9	3.00000
10	3.16227
11	3.31662
12	3.46410
13	3.60555
14	3.74165
15	3.87298
16	4.00000
17	4.12310
18	4.24264
19	4.35889
20	4.47213
21	4.58257
22	4.69041
23	4.79583
24	4.89897
25	5.00000
26	5.09901
27	5.19615
28	5.29150
29	5.38516
30	5.47722
31	5.56776
32	5.65685
33	5.74456
34	5.83095
35	5.91607

FIGURE B.41 Square roots of numbers.

FIGURE B.41 Square roots of numbers *continued.*

Number	Square root
36	6.00000
37	6.08276
38	6.16441
39	6.24499
40	6.32455
41	6.40312
42	6.48074
43	6.55743
44	6.63324
45	6.70820
46	6.78233
47	6.85565
48	6.92820
49	7.00000
50	7.07106
51	7.14142
52	7.21110
53	7.28010
54	7.34846
55	7.41619
56	7.48331
57	7.54983
58	7.61577
59	7.68114
60	7.74596
61	7.81024
62	7.87400
63	7.93725
64	8.00000
65	8.06225
66	8.12403
67	8.18535
68	8.24621
69	8.30662
70	8.36660

FIGURE B.41 Square roots of numbers *continued.*

Number	Square root
71	8.42614
72	8.48528
73	8.54400
74	8.66025
75	8.71779
76	8.77496
77	8.83176
78	8.88819
79	8.94427
80	9.00000
81	9.05538
82	9.11043
83	9.11043
84	9.16515
85	9.21954
86	9.27361
87	9.32737
88	9.38083
89	9.43398
90	9.48683
91	9.53939
92	9.59166
93	9.64365
94	9.69535
95	9.74679
96	9.79795
97	9.84885
98	9.89949
99	9.94987
100	10.00000

Number	Cube
1	1
2	8
3	27
4	64
5	125
6	216
7	343
8	512
9	729
10	1000
11	1331
12	1728
13	2197
14	2477
15	3375
16	4096
17	4913
18	5832
19	6859
20	8000
21	9621
22	10648
23	12167
24	13824
25	15625
26	17576
27	19683
28	21952
29	24389
30	27000
31	29791
32	32768
33	35937
34	39304
35	42875

FIGURE B.42 Cubes of numbers.

FIGURE B.42 Cubes of numbers *continued.*

Number	Cube
36	46656
37	50653
38	54872
39	59319
40	64000
41	68921
42	74088
43	79507
44	85184
45	91125
46	97336
47	103823
48	110592
49	117649
50	125000
51	132651
52	140608
53	148877
54	157464
55	166375
56	175616
57	185193
58	195112
59	205379
60	216000
61	226981
62	238328
63	250047
64	262144
65	274625
66	287496
67	300763
68	314432
69	328500
70	343000

FIGURE B.42 Cubes of numbers *continued.*

Number	Cube
71	357911
72	373248
73	389017
74	405224
75	421875
76	438976
77	456533
78	474552
79	493039
80	512000
81	531441
82	551368
83	571787
84	592704
85	614125
86	636056
87	658503
88	681472
89	704969
90	729000
91	753571
92	778688
93	804357
94	830584
95	857375
96	884736
97	912673
98	941192
99	970299
100	1000000

Diameter	Area
⅛	0.0123
¼	0.0491
⅜	0.1104
½	0.1963
⅝	0.3068
¾	0.4418
⅞	0.6013
1	0.7854
1⅛	0.9940
1¼	1.227
1⅜	1.484
1½	1.767
1⅝	2.073
1¾	2.405
1⅞	2.761
2	3.141
2¼	3.976
2½	4.908
2¾	5.939
3	7.068
3¼	8.295
3½	9.621
3¾	11.044
4	12.566
4½	15.904
5	19.635
5½	23.758
6	28.274
6½	33.183
7	38.484
7½	44.178
8	50.265
8½	56.745
9	63.617
9½	70.882

FIGURE B.43 Area of a circle.

FIGURE B.43 Area of a circle *continued.*

Diameter	Area
10½	86.59
11	95.03
11½	103.86
12	113.09
12½	122.71
13	132.73
13½	143.13
14	153.93
14½	165.13
15	176.71
15½	188.69
16	201.06
16½	213.82
17	226.98
17½	240.52
18	254.46
18½	268.80
19	283.52
19½	298.6
20	314.16
20½	330.06
21	346.36
21½	363.05
22	380.13
22½	397.60
23	415.47
23½	433.73
24	452.39
24½	471.43
25	490.87
26	530.93
27	572.55
28	615.75
29	660.52
30	706.89

Diameter	Circumference
⅛	.3927
¼	.7854
⅜	1.178
½	1.570
⅝	1.963
¾	2.356
⅞	3.748
1	3.141
1⅛	3.534
1¼	3.927
1⅜	4.319
1½	4.712
1⅝	5.105
1¾	5.497
1⅞	5.890
2	6.283
2¼	7.068
2½	7.854
2¾	8.639
3	9.424
3¼	10.21
3½	10.99
3¾	11.78
4	12.56
4½	14.13
5	15.70
5½	17.27
6	18.84
6½	20.42
7	21.99
7½	23.56
8	25.13
8½	26.70
9	28.27
9½	29.84
10	31.41

FIGURE B.44 Circumference of a circle.

FIGURE B.44 Circumference of a circle *continued.*

Diameter	Circumference
10½	32.98
11	34.55
11½	36.12
12	37.69
12½	39.27
13	40.84
13½	42.41
14	43.98
14½	45.55
15	47.12
15½	48.69
16	50.26
16½	51.83
17	53.40
17½	54.97
18	56.54
18½	58.11
19	59.69
19½	61.26
20	62.83
20½	64.40
21	65.97
21½	67.54
22	69.11
22½	70.68
23	72.25
23½	73.82
24	75.39
24½	76.96
25	78.54
26	81.68
27	84.82
28	87.96
29	91.10
30	94.24

Decimal equivalent	Millimeters
.0625	1.59
.1250	3.18
.1875	4.76
.2500	6.35
.3125	7.94
.3750	9.52
.4375	11.11
.5000	12.70
.5625	14.29
.6250	15.87
.6875	17.46
.7500	19.05
.8125	20.64
.8750	22.22
.9375	23.81
1.000	25.40

FIGURE B.45 Decimals to millimeters.

Inches	Decimal of a foot
⅛	.01042
¼	.02083
⅜	.03125
½	.04167
⅝	.05208
¾	.06250
⅞	.07291
1	08333
1⅛	.09375
1¼	.10417
1⅜	.11458
1½	.12500
1⅝	.13542
1¾	.14583

FIGURE B.46 Inches converted to decimals of feet.

FIGURE B.46 Inches converted to decimals of feet *continued.*

Inches	Decimal of a foot
1⅞	.15625
2	.16666
2⅛	.17708
2¼	.18750
2⅜	.19792
2½	.20833
2⅝	.21875
2¾	.22917
2⅞	.23959
3	.25000

Note: To change inches to decimals of a foot, divide by 12. To change decimals of a foot to inches, multiply by 12.

Fractions	Decimal equivalent
⅟₁₆	.0625
⅛	.1250
³⁄₁₆	.1875
¼	.2500
⁵⁄₁₆	.3125
⅜	.3750
⁷⁄₁₆	.4375
½	.5000
⁹⁄₁₆	.5625
⅝	.6250
¹¹⁄₁₆	.6875
¾	.7500
¹³⁄₁₆	.8125
⅞	.8750
¹⁵⁄₁₆	.9375
1	1.000

FIGURE B.47 Fractions to decimals.

Fraction	Decimal
1/64	.015625
1/32	.03125
3/64	.046875
1/20	.05
1/16	.0625
1/13	.0769
5/64	.078125
1/12	.0833
1/11	.0909
3/32	.09375
1/10	.10
7/64	.109375
1/9	.111
1/8	.125
9/64	.140625
1/7	.1429
5/32	.15625
1/6	.1667
11/64	.171875
3/16	.1875
1/5	.2
13/64	.203125
7/32	.21875
15/64	.234375
1/4	.25
17/64	.265625
9/32	.28125
19/64	.296875
5/16	.3125
21/64	.328125
1/3	.333
11/32	.34375
23/64	.359375
3/8	.375
25/64	.390625

FIGURE B.48 Decimal equivalents of fractions.

FIGURE B.48 Decimal equivalents of fractions *continued.*

Fraction	Decimal
$13/32$	.40625
$27/64$	.421875
$7/16$	.4375
$29/64$	.453125
$15/32$	.46875
$31/64$	.484375
$1/2$	.5
$33/64$	.515625
$17/32$	.53125
$35/64$	.546875
$9/16$	.5625
$37/64$	.578125
$19/32$	.59375
$39/64$	.609375
$5/8$	.625
$41/64$	.640625
$21/32$	.65625
$43/64$	.671875
$11/16$	.6875
$45/64$	.703125

Minutes	Decimal of a degree
1	.0166
2	.0333
3	.0500
4	.0666
5	.0833
6	.1000
7	.1166
8	.1333
9	.1500

FIGURE B.49 Minutes converted to decimal of a degree.

FIGURE B.49 Minutes converted to decimal of a degree *continued.*

Minutes	Decimal of a degree
10	.1666
11	.1833
12	.2000
13	.2166
14	.2333
15	.2500
16	.2666
17	.2833
18	.3000
19	.3166
20	.3333
21	.3500
22	.3666
23	.3833
24	.4000
25	.4166

Fraction	Decimal
1/32	.03125
1/16	.0625
3/32	.09375
1/8	.125
5/32	.15625
3/16	.1875
7/32	.21875
1/4	.25
9/32	.28125
5/16	.3125
11/32	.34375

FIGURE B.50 Decimal equivalents of an inch.

FIGURE B.50 Decimal equivalents of an inch *continued.*

Fraction	Decimal
⅜	.375
¹³⁄₃₂	.40625
⁷⁄₁₆	.4375
¹⁵⁄₃₂	.46875
½	.5
¹⁷⁄₃₂	.53125
⁹⁄₁₆	.5625
¹⁹⁄₃₂	.59375
⅝	.625
²¹⁄₃₂	.65625
¹¹⁄₁₆	.6875
²³⁄₃₂	.71875
¾	.75
²⁵⁄₃₂	.78125
¹³⁄₁₆	.8125
²⁷⁄₃₂	.84375
⅞	.875
²⁹⁄₃₂	.90625
¹⁵⁄₁₆	.9375
³¹⁄₃₂	.96875
1	1.000

Inches	Decimal of an inch
1/64	.015625
1/32	.03125
3/64	.046875
1/16	.0625
5/64	.078125
3/32	.09375
7/64	.109375
1/8	.125
9/64	.140625
5/32	.15625
11/64	.171875
3/16	.1875
12/64	.203125
7/32	.21875
15/64	.234375
1/4	.25
17/64	.265625
9/32	.28125
19/64	.296875
5/16	.3125

Note: To find the decimal equivalent of a fraction, divide the numerator by the denominator.

FIGURE B.51 Decimal equivalents of fractions of an inch.

APPENDIX C
OTHER USEFUL INFORMATION

Type of caulk	Relative cost	Life years
Oil	Low	1 to 3
Vinyl latex	Low	3 to 5
Acrylic latex	Low	5 to 10
Silicon acrylic latex	Medium	10 to 20
Butyl rubber	Medium	5 to 10
Polysulfide	Medium	20+
Polyurethane	High	20+
Silicone	High	20+
Urethane foam	High	10 to 20

FIGURE C.1 Characteristics of caulks.

- Gloss finishes are smoother and easier to clean.
- Gloss finishes are usually confined to use on exterior trim, while a flat finish is used on siding.
- The paint on a home exterior should last from 5 to 10 years.
- Exterior painting should not be done once the air temperature drops below 50°F. Extremely hot days are also days to avoid painting.
- Aluminum siding can be painted when the proper paint is used.
- With the exception of slate roofs and glazed tiles, all exterior surfaces are able to accept paint.
- Poor prep work and moisture are the two most common causes of peeling paint.
- Painting surfaces must be completely dry when working with oil-based materials.
- Most professionals paint windows after siding has been painted.
- Gutters and downspouts should be painted the same color as the siding is painted.

FIGURE C.2 Professional suggestions for exterior painting.

Tile	Rating
Ceramic tile	Fairly easy
Ceramic mosaic tile	Easy
Quarry tile	Fairly difficult

FIGURE C.3 Difficulty rating for installing tile.

- *Organic adhesive:* This product comes in a ready-mix mastic, applies easily, creates a flexible bond, and is available at low cost. Use of this material is limited to interior applications.
- *Dry-set mortar:* Created with a dry mix of cement, sand, and additives, this material resists freezing and immersion. However, the material must be kept moist for about 3 days prior to grouting.
- *Portland cement mortar:* Portland cement, sand, and water are mixed to create this setting agent. An advantage to this material is that it allows for minor adjustments in leveling the work area. However, tiles must be presoaked, and metal lath reinforcement is recommended.
- *Expoxy mortar:* Mixed by parts on the site, Expoxy mortar creates an extremely strong bond and is highly resistant to water and chemicals. Due to the nature of Expoxy, the time available for working with the quick-setting material is limited.

FIGURE C.4 Suggestions for tile setting materials.

Type of tile	Size of tile	Maximum joint width	Minimum joint width
Ceramic mosaic	$2\frac{3}{16}$" or less	$\frac{1}{8}$ inch	$\frac{1}{16}$ inch
Ceramic	$2\frac{3}{16}$" to $4\frac{1}{4}$"	$\frac{1}{4}$ inch	$\frac{1}{8}$ inch
Ceramic	$6" \times 6"$	$\frac{3}{8}$ inch	$\frac{1}{4}$ inch
Quarry	All sizes	$\frac{3}{8}$ inch	$\frac{3}{8}$ inch

FIGURE C.5 Recommended joint widths for floor tile.

Component	Material	Load (PSF)
Roofing	Softwood (per inch)	3
	Plywood (per inch)	3
	Foam insulation (per inch)	0.2
	Asphalt shingle	3
	Asphalt roll roofing	1
	Asphalt (built up)	6
	Wood shingle	3
	Copper	1
	Steel	2
	Slate (⅜")	12
	Roman tile	12
	Spanish tile	19

FIGURE C.6 Weights of building materials for roofing.

Component	Material	Load (PSF)
Framing (16" oc)	2-×-4 and 2-×-6	2
	2-×-8 and 2-×-10	3
Floor-ceiling	Softwood (per inch)	3
	Hardwood (per inch)	4
	Plywood (per inch)	3
	Concrete (per inch)	12
	Stone (per inch)	13
	Carpet	0.5
	Drywall (per inch)	5

FIGURE C.7 Weights of building materials for framing and floor.

Thickness	Width	Height
	Exterior	
1¾"	2'8" to 3'0"	6' 8" residential 7'0" commercial
	Interior	
1⅜"	2'6" min. bedroom 2'0" min. bath, closet	6'8"
Hardware		**Placement**
Door knob		36" above floor
Door hinges		11" above floor and 7" down from top of door Optional third hinge ½ way between other 2
Door clearance (Interior doors)		¹⁄₁₆" at top and latch side ¹⁄₃₂" at hinge side ⅜" at bottom

FIGURE C.8 Typical dimensions of doors.

Pipe size	Projected flow rate (gallons per minute)
½ inch	2 to 5
¾ inch	5 to 10
1 inch	10 to 20
1¼ inch	20 to 30
1½ inch	30 to 40

FIGURE C.9 Projected flow rates for various pipe sizes.

Pipe size	Number of gallons
¾ inch	2.8
1 inch	4.5
1¼ inch	7.8
1½ inch	11.5
2 inch	18

FIGURE C.10 Fluid volume of pipe contents for polybutylene pipe (computed on the number of gallons per 100 feet of pipe).

Pipe size	Number of gallons
1 inch	4.1
1¼ inch	6.4
1½ inch	9.2

FIGURE C.11 Fluid volume of pipe contents for copper pipe (computed on the number of gallons per 100 feet of pipe).

- Windows
- Doors
- Outside walls
- Partitions between heated and unheated space
- Ceilings
- Roofs
- Uninsulated wood floors between heated and unheated space
- Air infiltration through cracks in construction
- People in the building
- Lights in the building
- Appliances and equipment in the building

FIGURE C.12 Common sources of heat gain in buildings.

Material	Heat storage (Highest numbers are best)
Water	9
Wood	8
Oil	7
Air	6
Aluminum	5
Concrete	4
Glass	4
Steel	3
Lead	2

FIGURE C.13 Heat storage comparisons.

- Windows
- Doors
- Outside walls
- Partitions between heated and unheated space
- Ceilings
- Roofs
- Concrete floors
- Uninsulated wood floors between heated and unheated space
- Air infiltration through cracks in construction

FIGURE C.14　Common sources of heat loss in buildings.

- Bow
- Cup
- Crook
- Twist
- Check
- Split
- Shake
- Wane
- Knot
- Cross grain
- Decay
- Pitch pocket

FIGURE C.15　Types of lumber defects.

Application	Size	Notes
Siding	⅜ inch	
Wall sheathing	⁵⁄₁₆ inch	Studs 16" on center
Wall sheathing	⅜ inch	Studs 24" on center
Roof sheathing	⁵⁄₁₆ inch	Rafters 16" on center
Roof sheathing	⅜ inch	Rafters 24" on center

FIGURE C.16　Exterior use of plywood.

Usage	Live load (lbs./sq. ft.)
Bedrooms	30
Other rooms (residential)	40
Ceiling joists (no attic use)	5
Ceiling joists (light storage)	20
Ceiling joists (attic rooms)	30
Retail stores	75 to 100
Warehouses	125 to 250
School classrooms	40
Offices	80
Libraries	
Reading rooms	60
Book stacks	150
Auditoriums, gyms	100
Theater stage	150
Most corridors, lobbies, stairs, exits, fire escapes, etc. in public buildings	100

FIGURE C.17 Design loads.

Nominal size	Actual size	Board feet per linear foot	Linear feet per 1000 board feet
1 × 2	¾ × 1½	⅙ (.167)	6000
1 × 3	¾ × 2½	¼ (.250)	4000
1 × 4	¾ × 3½	⅓ (.333)	3000
1 × 6	¾ × 5½	½ (.500)	2000
1 × 8	¾ × 7¼	⅔ (.666)	1500
1 × 10	¾ × 9¼	⅚ (.833)	1200
1 × 12	¾ × 11¼	1 (1.0)	1000

FIGURE C.18 Board lumber measure.

Material	Weight in pounds per cubic inch	Weight in pounds per cubic foot
Aluminum	.093	160
Antimony	.2422	418
Brass	.303	524
Bronze	.320	552
Chromium	.2348	406
Copper	.323	558
Gold	.6975	1205
Iron (cast)	.260	450
Iron (wrought)	.2834	490
Lead	.4105	710
Manganese	.2679	463
Mercury	.491	849
Molybdenum	.309	534
Monel	.318	550
Platinum	.818	1413
Steel (mild)	.2816	490
Steel (stainless)	.277	484
Tin	.265	459
Titanium	.1278	221
Zinc	.258	446

FIGURE C.19 Weights of various materials.

Metal	Degrees F.
Aluminum	1200
Antimony	1150
Bismuth	500
Brass	1700/1850
Copper	1940
Cadmium	610
Iron (cast)	2300
Iron (wrought)	2900
Lead	620
Mercury	139
Steel	2500
Tin	446
Zinc (cast)	785

FIGURE C.20   Melting points of commercial metals.

Surface	Minimum	Maximum
Driveways in the north	1%	10%
Driveways in the south	1%	15%
Walks	1%	4%
Ramps		15%
Wheelchair ramps		8%
Patios	1%	2%

FIGURE C.21   Grades for traffic surface.

Material	Chemical symbol
Aluminum	AL
Antimony	Sb
Brass	..
Bronze	..
Chromium	Cr
Copper	Cu
Gold	Au
Iron (cast)	Fe
Iron (wrought)	Fe
Lead	Pb
Manganese	Mn
Mercury	Hg
Molybdenum	Mo
Monel	..
Platinum	Pt
Steel (mild)	Fe
Steel (stainless)	..
Tin	Sn
Titanium	Ti
Zinc	Zn

FIGURE C.22 Symbols for various materials.

Material	Expected life span
Aluminum	15 to 20 years
Vinyl	Indefinite
Steel	Less than 10 years
Copper	50 years
Wood	10 to 15 years

Note: All estimated life spans depend on installation procedure, maintenance, and climatic conditions.

FIGURE C.23 Potential life spans for gutters.

Material	Price range
Aluminum	Moderate
Vinyl	Expensive
Steel	Inexpensive
Copper	Very expensive
Wood	Moderate to expensive

Note: All estimated life spans depend on installation procedure, maintenance, and climatic conditions.

FIGURE C.24 Price ranges for gutters.

GPM	Liters/Minute
1	3.75
2	6.50
3	11.25
4	15.00
5	18.75
6	22.50
7	26.25
8	30.00
9	33.75
10	37.50

FIGURE C.25 Flow rate conversion from gallons per minute (GPM) to approximate liters per minute.

Vacuum in inches of mercury	Boiling point
29	76.62
28	99.93
27	114.22
26	124.77
25	133.22
24	140.31
23	146.45

FIGURE C.26 Boiling points of water at various pressures.

FIGURE C.26 Boiling points of water at various pressures *continued.*

Vacuum in inches of mercury	Boiling point
22	151.87
21	156.75
20	161.19
19	165.24
18	169.00
17	172.51
16	175.80
15	178.91
14	181.82
13	184.61
12	187.21
11	189.75
10	192.19
9	194.50
8	196.73
7	198.87
6	200.96
7	198.87
6	200.96
5	202.25
4	204.85
3	206.70
2	208.50
1	210.25

Activity	Normal use	Conservative use
Shower	25 (watering running)	4 (wet down, soap up, rinse off)
Tub bath	36 (full)	10 to 12 (minimal water level)
Dishwashing	50 (tap running)	5 (wash and rinse in sink)
Toilet flushing	5 to 7 (depends on tank size)	1½ to 3 (Water-Saver toilets or tank displacement bottles)
Automatic dishwasher	16 (full cycle)	7 (short cycle)
Washing machine	60 (full cycle, top water level)	27 (short cycle, minimal water level)
Washing hands	2 (tap running)	1 (full basin)
Brushing teeth	1 (tap running)	½ (wet and rinse briefly)

FIGURE C.27   Conserving water usage in gallons.

- A cubic foot of water contains 7½ gallons, 1728 cubic inches, and weights 62½ pounds.
- A gallon of water weighs 8⅓ pounds and contains 231 cubic inches.
- Water expands ¹⁄₂₃ of its volume when heated from 40° to 212°.
- The height of a column of water, equal to a pressure of 1 pound per square inch, is 2.31 feet.
- To find the pressure in pounds per square inch of a column of water, multiply the height of the column in feet by .434.
- The average pressure of the atmosphere is estimated at 14.7 pounds per square inch so that with a perfect vacuum it will sustain a column of water 34 feet high.
- The friction of water in pipes varies as the square of the velocity.
- To evaporate 1 cubic foot of water requires the consumption of 7½ pounds of ordinary coal or about 1 pound of coal to 1 gallon of water.
- A cubic inch of water evaporated at atmospheric pressure is converted into approximately 1 cubic foot of steam.

FIGURE C.28   Facts about water.

1. Wear safety equipment.
2. Observe all safety rules at the particular location.
3. Be aware of any potential dangers in the specific situation.
4. Keep tools in good condition.

FIGURE C.29 General safe working habits.

1. Do not wear clothing that can be ignited easily.
2. Do not wear loose clothing, wide sleeves, ties, or jewelry (bracelets or necklaces) that can become caught in a tool or otherwise interfere with work. This caution is especially important when working with electrical machinery.
3. Wear gloves to handle hot or cold pipes and fittings.
4. Wear heavy duty boots. Avoid wearing sneakers on the job. Nails can easily penetrate sneakers and can cause a serious injury (especially if the nail is rusty).
5. Always tie shoelaces. Loose shoelaces can easily cause you to fall, possibly leading to injury to yourself or other workers.
6. Wear a hard hat on construction sites to protect the head from falling objects.

FIGURE C.30 Safe dressing habits.

1. Read the operating instructions before starting to use the grinder.
2. Do not wear any loose clothing or jewelry.
3. Wear safety glasses or goggles.
4. Do not wear gloves while using the machine.
5. Shut the machine off promptly when you are finished using it.

FIGURE C.31 Safe operation of grinders.

1. Use the right tool for the job.
2. Read any instructions that come with the tool unless you are thoroughly familiar with its use.
3. Wipe and clean all tools after each use. If any other cleaning is necessary, do it periodically.
4. Keep tools in good condition. Chisels should be kept sharp and any mushroomed heads kept ground smooth; saw blades should be kept sharp; pipe wrenches should be kept free of debris and the teeth kept clean; etc.
5. Do not carry small tools in your pocket, especially when working on a ladder or scaffolding. If you should fall, the tools might penetrate your body and cause serious injury.

FIGURE C.32 Safe use of hand tools.

1. Be careful of underground utilities when digging.
2. Do not allow people to stand on the top edge of a ditch while workers are in the ditch.
3. Shore all trenches deeper than 4 feet.
4. When digging a trench, be sure to throw the dirt away from the ditch walls (2 feet or more).
5. Be careful to see that no water gets into the trench. Be especially careful in areas with a high water table. Water in a trench can easily undermine the trench walls and lead to a cave-in.
6. Never work in a trench alone.
7. Always have someone nearby—someone who can help you and locate additional help.
8. Always keep a ladder nearby so that you can exit the trench quickly if need be.
9. Be watchful at all times. Be aware of any potentially dangerous situations. Remember, even heavy truck traffic nearby can cause a cave-in.

FIGURE C.33 Rules for working safely in ditches or trenches.

1. Use a solid and level footing to set up the ladder.
2. Use a ladder in good condition; do not use one that needs repair.
3. Be sure step ladders are opened fully and locked.
4. When using an extension ladder, place it at least ¼ of its length away from the base of the building.
5. Tie an extension ladder to the building or other support to prevent it from falling or blowing down in high winds.
6. Extend a ladder at least 3 feet over the roof line.
7. Keep both hands free when climbing a ladder.
8. Do not carry tools in your pocket when climbing a ladder. (If you fall, the tools could cut into you and cause serious injury).
9. Use the ladder the way it should be used. For example, do not allow two people on a ladder designed for use by one person.
10. Keep the ladder and all its steps clean—free of grease, oil, mud, etc.—in order to avoid a fall and possible injury.

FIGURE C.34 Working safely on a ladder.

1. Do not lay tools or other materials on the floor of the scaffold. They can easily move and you could trip over them, or they might fall, hitting someone on the ground.
2. Do not move a scaffold while you are on it.
3. Always lock the wheels when the scaffold is positioned and you are using it.
4. Always keep the scaffold level to maintain a steady platform on which to work.
5. Take no shortcuts. Be watchful at all times and be prepared for any emergencies.

FIGURE C.35 Safety on rolling scaffolds.

1. Always use a three-prong plug with an electric tool.
2. Read all instructions concerning the use of the tool (unless you are thoroughly familiar with its use).
3. Make sure that all electrical equipment is properly grounded. Ground fault circuit interrupters (GFCI) are required by OSHA regulations in many situations.
4. Use proper sized extension cords. (Undersized wires can burn out a motor, cause damage to the equipment, and present a hazardous situation.)
5. Never run an extension cord through water or through any area where it can be cut, kinked, or run over by machinery.
6. Always hook up an extension cord to the equipment and then plug it into the main electrical outlet, not vice versa.
7. Coil up and store extension cords in a dry area.

FIGURE C.36 Safe use of electric tools.

1. Always keep fire extinguishers handy, and be sure that the extinguisher is full and that you know how to use it quickly.
2. Be sure to disconnect and bleed all hoses and regulators used in welding, brazing, soldering, etc.
3. Store cylinders of acetylene, propane, oxygen, and similar substances in an upright position in a well-vented area.
4. Operate all air acetylene, welding, soldering and related equipment according to the manufacturer's directions.
5. Do not use propane torches or other similar equipment near material that can easily catch fire.
6. Be careful at all times. Be prepared for the worst—and be ready to act.

FIGURE C.37 Preventing fires.

INDEX

ABOUT THE AUTHOR

R. Dodge Woodson is a plumbing contractor and licensed master plumber with more than 20 years' experience, and the author of many best-selling McGraw-Hill books, including *Home Plumbing Illustrated*, *National Plumbing Codes Handbook*, 2/e, *The Plumbing Apprentice Handbook*, *Plumbing Contractor: Start and Run a Money-Making Business*, and *Plumber's Troubleshooting Guide*. He currently teaches plumbing code and apprentice classes at Central Maine Technical College.